AUTISM

&

EVOLUTION

Spirit, Science & Fairy Tales

Marg Kinneen

 A catalogue record for this book is available from the National Library of Australia

Cover Image: **Sulada Jaichum** ©
Cover Design: Rachel Taylor © racheltaylor.com.au

Copyright © Marg Kinneen 2025
All rights reserved.
ISBN-13: 978-1-923174-53-5

Linellen Press
265 Boomerang Road
Oldbury, Western Australia
www.linellenpress.com.au

Contents

Contents iii

Introduction 3

Chapter One: Change The World 7

Chapter Two: Evolution 13

Chapter Three: Emotions 24

Chapter Four: Rethinking the Brain 41

Chapter Five: A Surprising Twist 62

Chapter Six: The Pineal and Pinocchio 79

Chapter Seven: The Light at The End of The Tunnel 104

Chapter Eight: Snow White and The Skeleton 125

Chapter Nine: Story 145

Chapter Ten: Denouement 171

Addendum: The Facts Unmasked 188

Acknowledgements 204

Word-Magic Glossary 206

Bibliography 253

"The famous scientist, Albert Einstein, who transformed scientific perspective in the beginning of the last century, also emphasised the unity of the universe and humanity.
He wrote:

> *A human being is a part of the whole called by us "the universe", a part limited in time and space.*
>
> *He experiences himself, his thoughts and feelings, as something separate from the rest – a kind of optical illusion of consciousness.*
>
> *This delusion is a kind of prison for us, restricting us to our personal desires and affection for a few persons nearest to us.*
>
> *Our task must be to free ourselves from this prison by widening the circle of understanding and compassion to embrace all living creatures and the whole of nature in its beauty."*

Albert Einstein (1879-1955), cited in
Consciousness and The Third Eye, Virendra Singh (2012)

Introduction

I started this book two years after the Asperger's diagnosis that tipped my world upside down and shattered my identity. I had no idea it would lead me to the understanding that there's nothing wrong with me.

It's my fervent hope that this book will help *you* to understand that there's nothing wrong with autistics and that those bearing the autistic label will see themselves through a new, and beautiful, lens.

When seen in the light of evolution, autism makes a great deal of sense. Indeed, I believe that autistics are the forerunners of evolution; that autism is not a disability, but is, in fact, a higher ability that's simply misunderstood.

Autism & Evolution presents a unique perspective that pushes the boundaries of accepted beliefs, including the current medical model of autism.

Evolution, by its very nature, is a boundary-pusher. What's considered normal shifts with the tides, although it forever lingers in the belly of the bell-curve, walking at the heels of those who break through the barriers of limitation and dare to imagine what's ahead of the curve.

So, be sure to engage your open imaginations, for there are many ahead-of-the-curve offerings in this book.

I straddle the arenas of Spirit and Science, so you'll see words like cells, DNA, the limbic system and hormones mingling with beliefs, chakras, fear and love, as I explore the exquisite plan of evolution that's bringing liberation to autistics; and, indeed, to us all.

In a delightful twist, you'll also encounter some of your favourite fairy tale characters as they step in to deliver information with simplicity and light. There's a wealth of profound hidden meanings woven within their stories.

I initially employed the magical Disney fairy tale versions that nestle in our hearts from childhood. However, as permission wasn't granted, I've reverted to the original fairy tales, which are less familiar. It's not the same, but I know that your wonderful imaginations will still conjure the magic and animate your inner spirits.

Another surprising revelation sees the Bible embrace evolution.

I see patterns and join dots across many genres. That's my gift. I'm not a pattern-*seeker*; I'm simply very observant and have a sense of curiosity and wonder about life. I feel certain that you, too, will see the patterns as I present my findings in this book. You may even wonder how you missed them previously.

Having said that, there will be those who don't align with my views. And I respect that.

However, *Autism & Evolution* is my understanding; my interpretation; my perspective. And it's precious to *me*. So, what's important is not that you believe what I say, but that I had the courage to say it. What's important is that everyone feels safe when expressing their creative truth; when publicly finding their voice, no matter how unusual it may appear.

My Glossary, which is situated at the back of the book, is most certainly unusual, although essential. It's introduced to the reader in Chapter Two, when I explain my special gift with individual words. Although I've referred to established dictionary meanings in some of the Glossary definitions, for the most part, this is not so. Therefore, don't expect the familiar, but do expect to be challenged. There are readers who've been so intrigued by the Glossary that they've read it from beginning to end even before they read the book. I'm sure there'll also be those who won't be interested in reading any of the Glossary entries. We're all different.

I do repeat some of the glossary references as we move through the text. This is always simply a suggestion in case you wish to revisit that particular definition to refresh your understanding as it applies to the current information. But the choice to do so is ultimately yours.

Words are far more powerful than we can imagine. They settle into our bodies, leaving subtle messages that shape our physical design. There's truth in the saying 'I feel it in my bones'. I'm guided by words. I feel their power and appreciate their service to communication. But communication can be derailed if the reader and the writer are on different tracks. So, I use *italics*. It's a choice I make if I'm using a word in a slightly different way, or if I'm employing a pun or double-entendre. And yes, there are times I simply use italics for emphasis. But I want you to know that this is *not* a slight on your intelligence.

Within the Glossary, I've devised a system to show the word break-downs as clearly as possible. This system uses capital letters, only because they make it clear. Please know that I'm *not* shouting at you. I would never do that.

Just as I would never disrespect you. When I use the word 'neurotypical' it's not a personal affront. I'm simply using a common term that's been adopted by the majority, including medical professionals, when referring to "people who have brains that function in a similar way to most of their peers." [www.medicalnewstoday.com *What does neurotypical, neurodivergent, and neurodiverse mean?*] This, by the way, is the first definition to pop up in a Google search for 'neurotypical'. It's followed by the definition from the Oxford Languages dictionary: "neurotypical: adj. not displaying or characterised by autistic or other neurological atypical patterns of thought or behaviour".

In employing the term neurotypical, I hope to demonstrate that labels can be vehicles of distortion, permanently branding individuals. Although in reality, labels are just perceptions that define the 'perceiver', not those being labelled.

Likewise, I wanted to address the many misperceptions about autism itself, so I've created an Addendum for this specific purpose. It's situated after Chapter Ten. Sometimes, all that's required is a different point of view.

This book isn't just a discussion about Autism and how it fits into the evolutionary progression. It also offers an opportunity to open your heart to the wonders within your own body, and to live your life with just a little bit more magic. After all, *everyone* is on the evolutionary spectrum into spiritual wellbeing.

In conclusion, I'm very aware that each autistic is unique and presents with remarkable variation. Autism itself is often accompanied by other conditions (co-morbidities...a term that badly needs refining), which colour the presentation of the autism. However, I firmly believe that all autistics sit under the same big-picture umbrella, and that, with a tilt of the lens, my unique perspective applies in every case. Please don't dismiss it because it's new, especially when the world is in need of alternatives.

So, I invite you now into the pages of my happy alternative, where autism fits perfectly within the natural progression of evolution.

The hardest challenge is to be yourself in a world where everyone is trying to make you be somebody else.

E.E. Cummings, *American poet*

Chapter One: Change The World

Great spirits have always encountered violent opposition from mediocre minds.

Albert Einstein

I used to believe that I didn't fit into this world. But now I realise that this *world* doesn't fit!

It's dragging its leaden feet, reluctant to recognise the natural course of evolution that's transforming us from mediocrity into…genius? From neurotypical to…autistic?

The accepted neurotypical perspective sees autism as a disability. My autistic perspective doesn't. Sadly, my perspective is overlooked because a radical alternative can be very scary. But so too can being autistic in a neurotypical world.

At first, I kept my Asperger's diagnosis a secret, because it was easier to hide it than to explain it.

I'm *different*!

But now, I wouldn't have it any other way. For woven beautifully within those nine little letters are the unique threads of my creative essence. My whole world is fashioned around my difference. My very purpose for being here is dependent upon my difference.

I still live in a world that doesn't fit *me*. But I've decided to use my difference to change the world. It's time to challenge neurotypical beliefs about autism. And from my front-row seat, I feel well equipped to do so.

When tip-toeing through the delicate arena of personal beliefs, we find that there are as many unique viewpoints as there are people on the planet. Each one perfectly valid, and providing the window through which the vista of an individual life is animated. The window of perspective creates one's world.

So, to change a perspective means to change your world. A big ask. But in so doing, you also change *the world*.

And that's my endeavour within the pages of this book. To invite you to see the world through my eyes; to take a peek through the window of my autistic perspective. It may unsettle, shock, or rock your boat. Perhaps it'll amuse. My hope is that it will pique your curiosity, and entice you to throw open the time-worn shutters of your carefully hewn window of perspective, and delight in the picturesque wonders beyond. I'm sure you'll find that it's really not so scary.

Let's change the world together...

Clearly, *the world* that we're about to change is the vista as seen through the eyes of each individual.

And yet somehow humanity has settled on a collective understanding, which creates the illusion that we're all seeing through the same lens. Unfortunately, this satisfies a sleepy majority, and is founded on fear; the fear of change; the fear of difference. But, apparently, majority rules!

> *Whenever you find yourself on the side of the*
> *majority, it is time to pause and reflect.*
>
> Mark Twain's notebook (1935)

What's deemed *normal* through the collective eyes of the majority, provides us with a familiar world of mediocrity. There's safety in numbers, so we define our world, and what's normal and acceptable, within the safe boundaries of popularity.

Dare to stand alone at your own peril. Mediocrity doesn't tolerate uniqueness; difference. An out-of-the-box world view is threatening to the status quo.

But it doesn't have to be. Stepping outside the box is always creative; it means opening to new possibilities and embracing new experiences; finding greater understanding and increased harmony. Can you imagine this out-of-the-box world?

History is full of creative dreamers who could; they leapt courageously out of their ill-fitting proverbial coffer, and paved a glorious path of ingenuity right up to your doorstep. And now, standing once again on the threshold of genius, we find the next wave of creative dreamers, called autistics.

Can I ask you to please park your reservations about the word genius for the moment, as all will become clear.

This world, as a collective mediocre box, is way too restrictive for those with autism. Indeed, my own wings feel tethered.

My world pushes the boundaries of possibility and understanding beyond what's currently accepted in science. I find patterns in everything, and join dots in obscure and inventive ways. Through the window of my perspective, I see a world where genius is the norm. I see this as our destination, the reason for our existence. I see the big picture of evolution.

This is what makes me different!

But autistics, by their very nature, see what others cannot.

Why?

Because autistics are the fore-runners of evolution; they're ahead of the pack on the continuum of time that's leading us into genius; they're preparing the way for those who follow, just as their ancestors prepared the way for them.

How can this *not* be obvious, especially with the increasing prevalence of autism. Do you really believe that our incredibly magnificent universe is leading us into an abyss of disability? I doubt it.

It's time to look at another alternative, and change *the world* view, not only about autism, but about why we're *all* here.

Our nurturing universe is taking us by the hand and leading us, generation after generation, into the dizzy heights of genius. Never before, in the history of the world, have so many enlightened souls graced our planet.

Now that's a big statement.

But with the understanding of *the world* being an individual perspective, it becomes clear that an increasing number of individuals are looking through a brighter lens and casting an enlightened perception upon the blue screen of planet Earth.

Dotted throughout evolution there have been enlightened souls whose pioneering spirits have captured the curiosity of seekers, and have opened people's imaginations to new possibilities. They've struggled against the religious and political norm, and been persecuted for their vision.

Their names are remembered in the annals of history as prophets, seers, masters, creatives. From Jesus, Buddha, Gandhi, Rumi, Da Vinci, Shakespeare and the classic poets, to the impressionist artists of the 19th century, Einstein and Hawking; they all share the crown of genius.

But I want you to know that this radiant crown lay dormant within us all, both literally and figuratively, awaiting the magic touch of inspiration. For genius is simply our guiding spirit of creativity and uniqueness soaring on the wings of belief.

And belief surely was the inner motivator of our brilliant predecessors. Self-belief fueled their passions and unleashed their unique, rare, quirky, and creative messages on hungry ears and eyes that were thirsting for something different — a new truth.

Stepping boldly through evolution, this avant-garde inclination effected a progressively wider influence on individual perceptions...and *the world* changed.

From this oblique angle it becomes obvious that the gradual unfolding of genius is infectious, and is indeed becoming epidemic, as evidenced by the increasing waves of creative genius...and autism...within the twentieth and twenty first centuries.

However, mediocrity cannot embrace a vision that hails genius as the norm, for that would put an end to itself. And so those enlightened souls who fall into the genius category of unique creatives are mistakenly proclaimed, by the blind majority, to be misguided or disabled. There must be something wrong with them if they don't fit the paradigm of mediocrity.

Autism being a case in point.

The deafness of the world is only selective insofar as mediocrity has a limited choice of what it's capable of believing. But the walls are closing in on the tight little box of mediocrity, which is slowly but surely shrinking in the evolutionary progression into genius.

We are indeed climbing out of mediocrity toward enlightenment. For evolution is simply a stroll through time that's leading us out of the shadows of ignorance, and down the path of change, into the glorious light of a new understanding.

Those enlightened souls who appeared to be rare creatures with special gifts were simply human beings who'd set their spirit free. In so doing, they dangled the carrot of opportunity for those who were to follow...us.

Unique creatives swallowed their bait, spread their eager wings, and took flights of fancy into the field of imagination and unlimited possibility. They dragged *the world*, kicking and screaming, into modernity, *because* of their unique perspective; *because* they were different. Let the scales fall from the eyes of mediocrity, and unshackle genius. Let the light of a new understanding liberate autistics from the neurotypical blind spot. Evolution is showing us that we're all rare creatures with special gifts. It's time to reveal the truth.

Everybody is a genius.

But if you judge a fish by its ability to climb a tree, it will live its whole life believing that it is stupid.

Albert Einstein

Chapter Two: Evolution

Sometimes it is the people no-one imagines anything of, who do the things that no-one can imagine.

Alan Turing from the movie The Imitation Game *(2014)*

In this chapter, I'm going to re-imagine evolution. And in so doing, I'll probably push the boundaries of *your* imagination.

The most obvious starting place is to ensure that we have the same understanding of the word evolution.

When researching the true meaning of any word, I like to visit a dictionary of etymology, for it traces the history of that word right back to its original, and sometimes obsolete, beginning; the source of its creation. The word evolution is intriguing:

> *… the opening of what was rolled up, to unfold, to make clear, to develop by natural processes to a higher state.*
>
> *Online Dictionary of Etymology*

It's my belief that autism provides evidence of evolution; that autism *is* the higher state that's developing through the progressive refinement of genes from generation to generation. Step by nurturing step, the flower of creativity is blossoming as the tightly rolled script of DNA reluctantly releases its grip on your imaginations. It's hard to surrender long-held habits, but autistics bear the proof of this unfolding.

Your wondrous imaginations are patiently poised within DNA, waiting for your eyes to open and liberate them from the religious exactitude of conventional custom...their debilitating adversary.

What's clear to me is that DNA is refining its cargo throughout evolution.

DNA is a double helix, with the original meaning of the word helix being *roll*. By applying one of the above definitions, we find that evolution is simply the un*rolling* of the long-held restrictive beliefs that are inherent in the familial scriptures of your DNA. Through the process of evolution, the tight bud of DNA is unfurling and releasing creativity from the clutches of tradition. Slowly, but surely, genius is blooming in the cellular garden of humanity.

The sole purpose for humanity's cyclical metamorphosis is to teasingly unveil the truth about creation, fostering a timely and gradual revelation that you are the creators of your reality; that the source of creation is your imagination; that it's powered by your beliefs; and that DNA is simply the vehicle through which evolution transpires. The fact that this truth is mysteriously hiding beneath the cloak of chemistry within your DNA confirms that evolution is taking place within your cells and that the fully animated book of evolution is transpiring within the human body of each and every individual.

The story you tell yourself about who you are is written on the parchment of your body tissue. Experience after experience shapes the narrative of your life into the physical manifestation of your body structure. The beliefs that you nurture are gloriously published, for all the world to see, resting in the glow of your skin, the curve of your bones and the light of spirit that dances in your eyes.

Each new generation epitomises an exhilarating new chapter within the cellular narrative of humanity. A gladdening of spirit. Excitement builds as the evolutionary plot pushes defiantly into new territory, flinging aside the old and courageously embracing the fresh breath of innovation that weaves its magic into the threads of humanity's body tissue, endorsing the vibrancy of youth and novelty. Evolution loves innovation, and so too does your body tissue.

Interestingly, the words novel, novelty and innovation hail from the same Latin derivative, meaning *new*.

Innovation f. into + Latin *novus*, new. Meaning "a novel change, experimental variation, new thing introduced into an established arrangement".

Novel f. Latin *novus*, new. Meaning "long prose fiction narrative or tale.

Online Dictionary of Etymology

The novel of Evolution, therefore, is a long-running cellular narrative, whose script is perpetuated via the relay of DNA; chapters forming as the innovations of each new generation are introduced into its established arrangement; changes written on the transformational body parchment of humanity, a published record of the creative ingenuity of an emerging imagination.

Evolution is a novel. A lovely little pun! For evolution means *change*, and novel means *new*. Clearly, anything that changes *is* new.

The fact that the novel of evolution is a fictional tale, reveals that the stories you tell yourself are simply your very own life fairy tales that satisfy the immediate needs of your generation; that these stories will be superseded by the higher imagination of the next generation; that they're a fanciful collection of experiences that are destined for the cumulative literary graveyard of history. But also, that in burying the *old* stories, you're daring to imagine yourself within the big-picture Happily Ever After of evolution.

Pushing the boundaries of the imagination always inspires innovation and moves the novel plot forward. It is innovation that animates the heart of autistics, who are unwittingly the great storytellers of evolution, as they narrate from a creative field that far exceeds the collective world view of mediocrity. As forerunners of evolution, their long ancestral trail of DNA has deposited them into the portal of their wide-open imaginations. And the physiological boundaries they've pushed in order to arrive here are the borders of their brain's limbic system. Not surprisingly, it's the limbic system that is the area of difference within autistics, which is erroneously seen to be underfunctioning. I'll fully elaborate on the limbic system, in relation to both autism and evolution, in Chapter Four (not to be missed). But, suffice it to say that walking the path of the forerunner invites misunderstanding from the less innovative or evolved, who themselves cannot imagine the higher state of genius in autistics, because it's beyond them.

We've already seen that the world view of mediocrity finds it hard to celebrate autism, tending instead to rate it as a disability. We also know that mediocrity baulks at the idea of change, preferring to maintain the status quo, because it feels safer and easier. It is wary of innovation, and is clearly an obstacle to evolution. Mediocrity rides through evolution on your DNA double helix. It takes the form of your family code of conduct, tradition, culturally accepted norms and obedience. Mediocrity feeds on a diet of control and approval, and wears the badge of neurotypical.

Approval is bound within the strands of DNA. To be a valued member of the family, society, culture or the world requires that you follow the script laid out before you by your familial ancestors. You must prove yourself worthy in their eyes; mirror their perception. They're not self-centred and mean. They're just doing what's always been done. From the moment of conception, you're fashioned to fit into *the world* as prescribed by others. To follow the dictates of expectation becomes the norm because it's easier than rocking the boat and tipping the ancestors into the unfamiliar waters of change.

Fear is inherent in this system of control and approval. Fear sits alongside mediocrity in the saddle of DNA. The family, society, culture and the world fear change for they have an emotional attachment to their beliefs. Therefore, *fear* courses through the cells of humanity, as it holds tightly onto the familiarity of the past. Fear imprisons evolution, innovation and the imagination behind walls of emotion within your stagnating cells...and calls it memory.

The past experiences that are programmed into the memory system of the human brain, dictate your chemical reactivity or emotional responses, as well as the physical form of your body tissue. When the *automatic* motivation of this memory system is challenged, your emotional responses are evolved, along with your body tissue. This is precisely what's been transpiring throughout evolution. The fear-based memory system is being soothed as the inherent patterns of old are being released, thus making way for the rise of innovation. This progressive liberation of the creative spirit is bringing increased feelings of joy, which is registered within the cells as lighter emotions, and within the more relaxed body tissue as increased integrity. Lending itself nicely to the slowing down of the aging process and the manifestation of increased longevity throughout evolution. The resurgence of the creative spirit, of innovation, of the imagination, is directly responsible for the increasing life span of successive generations. Evolution is replacing memory and emotion with imagination and youth.

Interestingly, the limbic system, within which evolution is pushing the boundaries of the imagination, and which is the area of difference in the autistic brain, is also the domain of both memories and emotions. More on the offerings of the limbic system in Chapter Four.

The little box of mediocrity is diminishing as humanity slowly rises above the emotional ties that call for control and approval.

Far from just wanting to prove themselves in a neurotypical world, autistics want to spread their expansive wings of innovation, and lift the world out of the ordinary and normal, and into the wonders that grace their extraordinary imaginations.

But how is this possible when they're seen to be disabled?

It begs the question, where does the disability actually lie, when it's mediocrity that's *unable* to rise above its debilitating fears and emotions? Perhaps it's time for neuro-typicals to check their own perspectives, for they may find them lacking. Are you disabling *us*?

This isn't intended as a bald accusation of blame or guilt. It's simply an invitation to walk a mile in a pair of autistic shoes, to see if it changes your neuro-typical landscape.

Neurotypical, in itself, is an interesting turn of phrase.

Before dissecting these two little words, I'd like to share a secret with you. My personal, and rather unusual, gift is that I *see patterns in individual words, from which I join obscure dots to reveal deeper meanings*. This treasured lagniappe provides me with a unique understanding of evolution and how it's progressing within the physiology of our human body. It's been both a blessing and a curse.

I love how words speak to me and reveal their intimate secrets. And I can discourse on this information ad infinitum, it fascinates me so. But the double-edged sword brings me sharply back to earth bearing wounds of humiliation from those who don't understand, the acquiescent majority. With my sanity being questioned, I soon learned that it was safer to keep it hidden. Fortunately, I had a very progressive and understanding doctor, who provided me with the courage to see that I am, in fact, sane. I'm very grateful for his expanded vision, which has pulled me through many years that were fraught with frustration. Without his support and kindness, I wouldn't have embraced my uniqueness, or found my voice. That's all it takes...one person to believe in you. One safe space.

Although I was deliciously tempted to lace this text with examples of my word-magic, I'm acutely aware that it may leave you blankly befuddled. So, for those of you who *are* interested in this clandestine treasury of information, I've created an alphabetical Word Glossary at the back of this book, to which you can refer when directed, if you so desire. Here's an example using the word *nerve*. **[nerve: see glossary]** Now you can pop on over to the glossary, if you wish, before reading the next passage. But be prepared for your preconceptions to be challenged, for these are not your average dictionary definitions, but information channelled from a higher source. However, if you're not interested in the glossary, you can simply continue reading. For those reading an e-book, click on the little superscript number beside the word *nerve*, and a pop-up box will appear with the appropriate glossary entry. And please be aware that there may be a few occasions where a word breakdown appears within the body of the text, in order to bring increased understanding.

That said, let's dive into the inner sanctum of the term *neurotypical*.

Clearly, *'neuro'* relates to the nerves or nervous system, and *'typical'* means normal, ordinary, usual, or conventional. So, a neurotypical is someone who has normal electrical impulses firing through their nervous system, informing their brain and body. *Normal* messages dictate their beliefs, behaviours, body structure and functions. Convention is apparent within the neuro-typical physiology, as it conforms to an expected pattern or rule. So here we are back to mediocrity again as the ancestors direct the neural traffic in accordance with habit. Even within our physical bodies, majority rules.

But further to that, we find that *fear* is fundamental to the neuro-typical. In fact, fear is inherent within the nervous system itself. The nervous system *is* a system of fear…it's very *nervous*.

Electrical impulses carry these messages of fear to the brain, which in turn sets up the chemical defence system in the aforementioned memory of the neuro-typical. Surely a *defence* system is only necessary in a world where fear exists? **[defence: see glossary]**

The ancestors are toey! Teetering on the edge of a deathly precipice…the generation gap…for fear of being usurped by new ideas or perspectives.

Travelling through history, evolution has always been accompanied by a generation gap. It's never been easy for the current generation to embrace the new-fangled ways of the young. Misunderstanding and unrest have always preceded acceptance and resignation. But what if the gap becomes so wide, and the understandings so advanced, that the previous generations simply can't fathom the innovations or the individuals delivering them? It's beyond their imaginings. So, from a position of self-preservation and protection, the unfathomable is relegated to disability, and the arena of research and development. This, I believe, has been the fate of autism.

It's only the fear-of-the-unknown that prompts these actions, and the searching minds of neuro-typical scientists soon advance their understandings, as humanity leap-frogs its way from discovery to discovery. However, in the case of autism, it's going to take more than the neurotypical mind to fully traverse the gaping chasm of the current generational shift into genius. Because, dare I say it, it takes one to know one.

The neuro-typical is an individual whose worldview operates, subversively, on fear. But evolution is steadily overcoming this fear by disarming the defence system, thus altering the emotional behaviour or chemical reactivity within the cells. By taming your emotions, evolution is soothing and altering the nervous system. In fact, its ultimate mission is to render it obsolete. No more fear, or defence. Autism, seen as a neural developmental *dis*order, bodes well for evolution, with its reduced development of fear, upon which I will elaborate in Chapter Four, when I discuss the limbic system and its defence mechanisms.

Fear is a thing of the *past*...literally. It's entwined in the *memory* traces of your ancestors who ride the waves of mediocrity within your DNA. Your cellular memory is a chemical concoction of past emotions that's been handed down to you through the ages. Therefore, neurotypicals live in the past, holding on to old beliefs so as to gain approval from the languishing majority, in an endeavour to save themselves from the pain of retribution. In this, there is no blame; just an inbuilt design for self-preservation.

With the introduction of this label, neurotypicals may have inadvertently cast themselves in the least desirable role.

Thankfully, common sense tells us that it's the neurotypicals who are evolving in the novel called Evolution, as it progressively transforms the definition of *typical*. Each new generation establishes its unique typical identity. With the gradual deletion of obsolete ancestral memories, the past is continually being reimagined into an innovative present. Even so, it's going to take a good measure of open-hearted grace by some exceptional neuro-typical scientists to publicly recognise and allow themselves to be guided by the genius in autism.

> *When a true genius appears in the world, you may know him by this sign: that the dunces are all in a confederacy against him.*
>
> Jonathan Swift (1706)

> *Jealousy is the tribute mediocrity pays to genius.*
>
> Archbishop Fulton J. Sheen

In staying true to its definition, evolution is developing by natural processes, spreading its wings and soaring into the higher vibrational state of genius that rides the creative vibes of imagination. Brilliance flaunts its glory through radiant threads of body tissue infused with light, as it transmits the magnetic essence of love.

Humanity is evolving from fear into love, and autism is at the *heart* of this transformation. **[heart: see glossary]**

Evolution is quite literally a change of heart, for the cell nucleus, as the heart of every cell, is where this change occurs. It's here that your DNA has fought its patriarchal battles against innovation throughout its long evolutionary trek. And it's here that the last vestiges of your DNA are now poised to surrender peacefully, embraced by the loving arms of your nurturing *mother*-nature. (I'll be elaborating on your *mother*-nature, in relation to your immune system, in Chapter Four)

Peeking through the window of my perspective provides a seemingly topsy-turvy view that turns accepted scientific tenets on their neuro-typical head. For me, as an Aspie, it's a no-brainer! My hope is that your imagination is supple enough to entertain the *possibility* of something that's a little different, a little innovative, a little unique. For as the opening quote states,

> *Sometimes it's the people who no-one imagines anything of, who do the things that no one can imagine.*

Autistics, who the majority consider to be disabled, constantly surprise neurotypicals with their creative ingenuity!

With a little imagination, it can be clearly seen that autistics do fit in...to the *big* picture of evolution.

> *...encourage them to unfurl their wings - past any horizon, past even our wildest expectations.*

Kristin Barnett (2014) The Spark: A Mother's Story of Nurturing Genius and Autism. (page 250)

Chapter Three: Emotions

What might be taken for a precocious genius is the genius of childhood. When the child grows up, it disappears without a trace.

It took me four years to paint like Raphael, but a lifetime to paint like a child.

Pablo Picasso

Emotion: f. Latin *emovere* = *ex* <u>out</u> + *movere* <u>to move</u>, meaning "move out, remove; agitate".
Online Etymology Dictionary

Emotions appear to be on the move. But *what* exactly is moving? From where is it moving, and to where is it moving? My curiosity is piqued. Perhaps yours is too. Let's find some answers.

Emotions. You encounter them every day. You experience your own and those of others. They're a natural component of life, it would seem. Emotions frequent all your relationships; at home, work, sports and hobbies; and your favourite books and movies...what would they be like without their emotional content?

Can you even imagine what the world would be like without its emotional highs and lows?

Well, evolution can. Evolution has been re-imagining a world without emotions throughout its duration, and dragging humanity out of the emotional clutches of mediocrity in its wake.

The little box of mediocrity is diminishing as humanity slowly untangles its emotional ties. Clearly, evolution has an emotional inclination. It's your emotions that are evolving within your cells, as beliefs are progressively refined through the macrocosmic cycle of generations. However, it's the individual who has the power to ultimately determine their own patterns of belief and emotion. Therefore, it's the individual who dictates the rate of evolution within their own cells. Only when you realise that evolution is occurring at an individual level, will you be able to embrace *difference*, and change your worldview for the better. Only then will it be clearly understood just how autism fits into the big picture of evolution and why it presents with such a diverse range of symptoms.

In this chapter, I'm going to address the tricky minefield that is emotions, reveal scientific evidence showing how they affect your physiology and reality, and explain the emotional proclivity of autistics. **[emotion 1 & 2: see glossary]**

The salty ocean of electro-chemistry within your cells is the electromagnetic ocean that carries your emotions, dictating your behaviours and radiating into the world around you to become one with the electromagnetic field. Suddenly, the electromagnetic light field is seen to be a living field of emotion.

So now we can see that it is electromagnetic radiation that's *moving out* of your cells and into the electromagnetic field of light. The process of emoting sees your energy moving out of its cellular home into the big wide world. My curiosity is sated. However, this also means that your body is effectively losing energy, which becomes apparent in the other description of the word emotion…agitate. According to the Oxford Dictionary of Etymology, *agitate* means 'to disturb, to put in constant or violent motion, drive onward, impel, incite to action, stir up'. The agitating nature of your emotions clearly expends a lot of energy, as they're thrust from your cells to mingle with the surrounding electromagnetic field. Ultimately your cells are losing *power* whenever you express an emotion.

There's an interesting connection between the words agitate, hormone and emotion, which confirms that they are the driving force within your cells. Perhaps not a glowing reference, as it emulates the controlling external influences that whip you into shape during childhood; the powers-that-be who ensure that you comply with expectation so as to fit into the world around you. Even the electromagnetic field is a fabrication of expectation.

The electromagnetic field is a big, fluid tapestry that's woven from the individual emotional threads of humanity. Each thread contributes to the masterpiece of reality. *Your* individual threads of emotion move the great tapestry of the electromagnetic field, affecting every other thread, and changing the overall design of reality. And in turn, your individual threads are affected by the movement of all the other threads. Emotion has power over reality in the external world! We're all connected through the emotionally charged electromagnetic field, which is why we pick up on people's emotions or the atmosphere in a room we've just entered.

Why is this so?

In 2017, after joining the dots from three separate experiments, scientists proclaimed, with great astonishment, that **human emotion physically shapes your reality**. Now, that's probably not so astounding a revelation for the more spiritually inclined, but it could be quite a leap for the die-hard scientist. Thankfully science is indeed beginning to open its eyes to the profound truth that's inherent in long-held spiritual understandings, slowly but surely establishing them as fact.

After extracting DNA from their subjects, and isolating it up to a distance of 50 kms, it was discovered that the DNA responded precisely to the emotions experienced by the subject, with negative emotions tightening the DNA coils and positive emotions relaxing them. It became clear that your emotions shape your DNA. In other words, you design your own body by virtue of your emotional proclivities. And that, in the beginning, you were created by a little package of ancestral emotions.

The second experiment also isolated subjects and their DNA to discover that the DNA exhibited the same electrical responses (emotional peaks and valleys) as the subject, *at exactly the same time*. So not only do your emotions communicate with, and affect, your genes, but they do so beyond space and time, thus defying conventional laws of physics.

The proverbial light bulb flashed when the third experiment revealed that the DNA affected light photons, which are the quantum particles of the electromagnetic field. The light photons precisely followed the geometry of the DNA they came into contact with. **[geometry, reality: see glossary]**

So now to join the dots. Your emotions affect your DNA, which in turn affects the light photons that make up your reality and the world around you (within the electromagnetic field). Therefore, your emotions physically affect the world, and any change of emotions physically alters, or evolves, that world. *You create your world...and shape evolution.*

> *Human emotion literally shapes the world around us. Not just our perception of the world, but reality itself.*

> *They concluded that human DNA literally shape the behavior of light photons that make up the world around us!*

Life Coach Code (20217), New Research Shocks Scientists: Human Emotion Physically Shapes Reality

Clearly, your cellular electromagnetic oceans affect and alter the collective electromagnetic field. Light photons in the electromagnetic field are particles of emotion. From the microcosmic cells to the macrocosmic physical world, emotions rule.

The electromagnetic field is essentially a mirror of light that reflects the emotional state of humanity, in an enormous feedback loop called reality.

Each individual human being contributes to the electromagnetic field of emotion that connects us all. By association, each individual human being contributes to the *evolution* of the world around them every time they change their emotions.

This tells us that evolution is measured on a continuum of emotions that reveals itself through the electromagnetic field of light, otherwise known as *the world*. The world, both in reality and your perception of it, is governed by your emotions. Accordingly, the electromagnetic spectrum contains the whole spectrum of emotions, from the heaviest to the lightest. Lower light frequencies carry heavier emotions, like sadness, grief and anger, creating increased dis-ease within your cells and your world reality. Therefore, the higher the light frequency, the greater the ease in your cells and your world reality.

What happens in your cells, manifests as your reality. But remember that your cells have no power of veto; they're extremely obedient; they simply receive your messages and express your emotions accordingly, just like a programmed computer.

Each and every individual sits along the spectrum of light, aligning with their particular emotional stage of evolution, which in itself can vacillate momentarily or periodically. But rest assured that collectively we're steadily ascending toward the radiant field of light that sits on the highest frequencies, creating constant ease within the cells of each individual, their physical body and the larger world reality. That is the natural course of evolution that's leading us out of mediocrity and into genius.

So, how does genius connect with emotion?

All electromagnetic light waves carry information in relation to their particular frequency. The more refined, higher light frequencies carry more light and energy, higher information, more expanded understandings and greater innovation. As they move toward the genius end of the spectrum, they dump their emotions and lighten their load. And when you connect fully with the most refined, universal field of light, where all information already resides, you not only open to universal knowledge, but also to the truth that only love exists. For you realise that anything less than love is simply an illusion that rides on the fear-filled lower vibes of emotion.

I know this is all sounding very spiritual. That's not my intention, but it *is* the natural inclination when recognising the characteristics of light and understanding its effects on the reality that is your physical body. There is an undeniable relationship between light frequencies and emotion. Just because it has a spiritual essence doesn't make it any less scientific. In fact, the energy that courses through your body tissue is your power, courtesy of your spirit. You're all spiritual beings expressing through a physical body. It's time to rescue the term spiritual from the clouds of woo-woo and ground it firmly within matter, for without it, you wouldn't exist. And evolution would cease.

Evolution is refining and overcoming emotions on its generational climb into higher light frequencies, fostering a resurrection of the physical body.

In fact, I'll go so far as to say that the genius light frequency is beyond emotion; that emotion only exists on light frequencies less than genius. When evolution crosses the genius finish line, the electromagnetic field will have transformed into a pure magnetic field of attraction. This is the universal field of creation, otherwise known as the field of imagination. And your imagination is the most powerful force in the universe. **[magnet & imagine it: see glossary]**

Every human being sits along the evolutionary spectrum of emotion. Autistics are simply situated at the higher frequency end, exhibiting fewer emotions. They're poised to ride the waves across the finish line of love and genius, which is evident in their gentle, nature-loving spirits, their acute sensitivities, their creativity and their brilliance.

Love and genius are synonymous. Autistics exhibit both.

It's important to understand that, as increased numbers of individuals are refining their beliefs and behaviours and moving into higher light frequencies through the evolutionary progression, the goal of genius seems less intimidating. We're all heading in that direction as we refine our emotions.

From a physiological perspective, emotion (cellular electromagnetic ocean) is governed by the neuro-endocrine system within the body. The electrical firings of the nervous system determine the chemical composition within the endocrine system, which exhibits your emotional reactions to the external world. We've already addressed the fact that the nervous system is a system of fear...it's very nervous. So, in an unexpected twist, the emotional system of the human body is governed by fear, as is the electromagnetic field. From the emotionally-driven human persona, we all connect and communicate through a dense field of fear, which we call the atmosphere. **[atmosphere & endocrine: see glossary]**

By association, the evolutionary journey through the spectrum of emotion into higher light frequencies, is a continual process of *overcoming* fear, which registers within the neuro-endocrine system of the human body.

Evolution is witness to mankind *coming up over* its lower programmed patterns of belief, behaviour, understandings, fears and emotions, thus calming the cellular oceans and weaving finer threads into the electromagnetic tapestry called the atmosphere. The collective blanket of emotions is becoming more refined, its filaments more radiant. And this gradual crawl of evolution is picking up momentum, as evidenced by the increase of highly sensitive, intelligent and loving autistics. Their neuro-endocrine systems bear witness.

Unfortunately, in the current autism climate, you'll encounter a heavy blanket of fear at every turn, for it is understandably the predominant and all-pervading energy that drapes itself over worried parents, therapists, educators and researchers. However, the fear *around* autism is, in itself, contributing to the illusion that an epidemic of disability is looming. The word autism is dripping with fear. But if you take away the notion of disability and replace it with the enlightened understanding that autistics are higher functioning, everything changes. By seeing autistics through a brighter lens, you allow them to follow the natural urges of their creative genius, and suddenly fear dissolves into celebration. With a huge sigh of relief, a mighty boost of loving energy surges through the cells of the autistic juggernaut, wrapping humanity in a much-needed nurturing hug of light. Evolution smiles as a long-awaited epidemic of genius obligingly emerges. Simply through a change of perspective.

In moving forward and discussing the role of family in the evolutionary pursuit, please understand that there is no blame intended, particularly in relation to autism. Families are amazing, courageous, exhausted, nurturing saviours for their autistic loved ones. I'm simply stepping inside the human body and taking a physiological perspective of family and how it settles within your body tissue. Please don't take it personally.

Physiologically speaking, family is simply a bundle of energetic characteristics, in the form of emotions, that's following the evolutionary path through the neuro-endocrine system of individuals.

The endocrine system of the human body is the partner in crime with the nervous system, fulfilling its fearful script by secreting family obligation and expectation within the very fibre of your being.

Once the *nervous* prescription has been filled by the brain, the emotional chemical concoction is delivered to the cells of your endocrine organs in the form of hormones, which then set your patterns of behaviour in motion (emotion) in an attempt to deal with the perceived threat. This perception of threat is founded upon the programmed data within your memory, which is based on all your learned responses to childhood family experiences. Family takes residence within the cells of your body, becoming the driving force.

Your endocrine system holds the secret fears of the family, and all their emotional patterns of reactivity associated with those fears, in the form of hormones. Your chemical secretions are the family secrets. During childhood, these fearful family doctrines are gathering into a collective bible of beliefs that bursts open upon puberty. The family secrets become public in a fireworks display of long-held emotions; your suppressed behaviours that have been kept in check because you didn't have a voice. Puberty pushes your new-found adolescence into the pulpit of life, where you begin the long journey back to reconnecting with, and liberating, your true essence. **[adolescence & hormone: see glossary]**

The *neuro-endocrine* system is a fear-guilt system of emotion, driven by control and culminating in suppressed anger, which bubbles away in your cells; little cauldrons of passion. This system of fear and guilt is founded upon the collective familiar beliefs that form the family code of conduct, dictating the habits that you practice religiously, which become your instinctive patterns of reactivity…your hormones. Your family's religious practice is embedded in your hormones.

Expectation has firmly planted this bible of *religious habits* into your obedient cells, *clothing* your body in a mantle of mediocrity. For expectation is a thief of creativity, and the arch-nemesis of evolution. Expectation gobbles up your uniqueness as the fear of impending punishment wields its implied warning. Your neuro-endocrine system is a system of expectation; a system that is guided by the external world of mediocrity with which you're programmed to comply. Your *inner* guiding system, the loving inner compass of your original creative spirit, has been replaced by a tangled system of fear and guilt that's designed to please the people in gestures of popularity. Fear is a people pleaser. Fear and mediocrity are allies. **[expectation: see glossary]**

Fear and mediocrity are allies, just as love and genius are allies.

But what of individuality? What of pleasing your genius and inspiring your unique essence?

Enter evolution. **[evolution 1, 2 & 3: see glossary]**

Evolution has been gently drawing the veil of illusion from the eyes of humanity, and liberating the creative spirit through a generational cleansing of outdated, fear-based belief patterns. Evolution is transforming humanity from the illusion of fear, which only exists in a lower belief system, and transporting it into a new reality of love. Whatever you believe becomes your reality. **[belief: see glossary]**

Evolution, thus far, has presented itself as a continuum of light, that's overcoming fear and emotion through the progressive refinement of self-belief.

You hold the reins of evolution, because you have the power to steer your own course, by choosing to believe in the highest possibilities for yourself, and your world. You do evolution a favour when you believe in your unique abilities and express them creatively. It's time to step out of the family fortification that's kept you trapped in the dungeon of dark, emotional cells, languishing in frustration and anger at the indignation of having to do someone else's bidding rather than follow our own heart. Have courage and stand in your truth, and shine your light, so others can do the same. Embrace your difference; dance with your uniqueness.

And throw the light of gratitude upon autism.

For autism is providing the opportunity to impact enormous change upon society and the world, by forcing the family to take a fresh look at their patterns of belief and behaviour. Autism is forcing education, therapy and research to take a fresh look at *their* patterns of belief and behaviour. Autism is challenging and changing the world view, simply because it doesn't fit this world pattern.

Evolution beckons a common-sense pattern that accommodates, and relishes, individual differences. Let's do away with the old measures that bundle everyone together in a scale of *normal* developmental progression, and focus instead on the amazing and unique gifts afforded to each individual, no matter how those gifts present themselves. Let's remove all expectation, and stop focusing on what people can't or shouldn't do, so we can celebrate what they love; what bathes their little cells in an incandescent glow, and radiates joyfully through their sparkling eyes. Let's find what's at the heart of each individual; what lights them up from within. I believe that evolution is creating a *world of difference*, where everyone fits in because uniqueness is celebrated. Autism is a game-changer.

Nowhere is this more apparent than in the short movie 'Life, Animated' based on the book written by journalist Ron Suskind, which tells the story of his son, Owen, who vanishes into autism around the age of three.

The movie opens with this telling quote by Owen himself. "There is a boy who is just like other boys, until one night he sees from his window, a storm on the horizon."

Owen is seeing, through the wide-eyed window of his perception, the impending storm that's inherent in growing up. This powerful analogy is fraught with fear, as the young Owen senses the weight of expectation that's marching toward him on the back of his future adult persona. Growing up presents a stormy assault on the magical world of the child. *[Life, Animated; A Story of Sidekicks, Heroes and Autism. 2014. Ron Suskind, journalist, author and filmmaker]*

To fully understand this onslaught, let's take a wide-eyed peek at what we leave behind in the process of growing up. **[infant 1 & 2: see glossary]**

Inherent within infants are delta brainwave patterns inspiring intuition, imagination and regeneration, which are characteristic of the higher power and higher light frequencies that course through their youthful little bodies. At embryonic infancy they're still connected to the universal power source of creation, the treasury of love, dreams and youth, that expresses through stem cells. *Intuition* is present in the form of loving inner guidance from a sparkling new and fully functioning belief system that's yet to be tested; *imagination* carries the power to attract unlimited possibilities, creativity and make believe; and *regeneration* lovingly nurtures their body tissue, thus promoting ongoing youth. The delta brainwave patterns of the embryonic infant inspire the radiance of the pregnant mother. **[delta: see glossary]**

This wondrous and magical land of uniqueness and possibility is already neatly tucked within the heart of every infant. A magical life fairy tale is poised to animate the world around them through the free expression of their creative spirit. Joy plays within the very fibres of their little bundles. This is the universal *Once Upon a Time* fairy tale-setting that's inherent within each and every human being that's graced the planet throughout evolution. Uniqueness is your inherent design. And it carries with it a lightness of spirit that's truly magnetic. Can you feel it? It's still within you. In fact, it's your first and fondest memory. And it's magical. So why would anyone want to leave such a place...or change it?

Owen certainly didn't want to. Instead, he firmly closed the gates of his fairy tale castle, and shut out the stormy evil villain that was descending upon his uniqueness. As an adult he said, "I've been scared my whole life of growing up. Peter Pan doesn't want to grow up, because when you grow up, you lose all your magical, childhood times." *[Life, Animated; A Story of Sidekicks, Heroes and Autism. 2014. Ron Suskind, journalist, author and filmmaker]*

Only an innocent misunderstanding would replace a magical world with a stormy one that's filled with fear. That innocent misunderstanding is the blind belief that children must follow the dictates of adults so they can grow up and fit into the external world...with the implication that adults know better than children, and that the external world is the best option. *Innocent* because there's no negative intent in following a fundamental belief that's considered normal. A *misunderstanding* because children are, in fact, more connected to the universal field of truth, possibilities and creation than adults.

Unfortunately, it's the genius of childhood that *moves out* of its original infant home into the adult world, carrying its baggage of emotion.

I believe that autistics are sending a silent and powerful message to the world, on behalf of all children throughout evolution. *Please don't take away my magic.* **[magic: see glossary]**

Through successive generations, children have followed the stormy dictates of expectation, albeit with increasing rebellion as they've been slowly finding their collective voice. However, the last vestiges of mediocrity are still clinging stubbornly to their old paradigms, refusing to hear this little voice of genius. Enter autism, an unorthodox solution, where children are withdrawing themselves, in increasing numbers, and finally getting the world's attention. Through this simple act of self-preservation, autistics are leading the evolutionary procession out of its stormy mediocrity by simply refusing to play the game.

This collective silent protest is not only evolving the belief patterns of individuals, nourishing them both spiritually and physically, but also fostering higher light frequencies in the electromagnetic field that cocoons humanity as a whole. Reality is evolving. The world view is opening to higher beliefs and the increased possibility of a more radiant future. Mediocrity is transforming into brilliance, one little light at a time, weaving a twinkling web of love through the tapestry of humanity. The tight little box of mediocrity is literally bursting...with light. A future of genius seems far more in keeping with this magical universe than one of disability. Just ask the children.

Autistics have chosen to withdraw from this dense world of emotion, expectation and rules. Yes, they do inhabit their own private world, but they don't disappear into a blank void. They're much happier in their own little world of genius, creativity and ease, because it's filled with all the wondrous things they love and believe in. They're just waiting for you to believe in them too. Open your eyes to the magic that adorns their high-frequency world, because it's sitting deep inside you too, waiting to be re-discovered, and you'll find that it's literally out of this world! Leave your mediocre world behind by opening to a system of higher beliefs that provides you with a new understanding about autism.

In his book *The Reason I Jump (2014)*, thirteen-year-old autistic, Naoki Higashida, states,

> *We just want to go back. To the distant, distant past. To a primeval era...we are a different kind of human, born with primeval senses...*

Primeval means *belonging to the first age*. Naoki wishes to return to the distant past; to his first age; to infancy. He, like all autistics, intimately feels the pull of his first and fondest memory that's sitting within the fibres of his being. What Naoki doesn't realise is that *everyone* is born with primeval senses, and everyone feels the pull to return to this happy home. Autistics are simply the first to arrive. And they've put out the welcome mat for the neurotypicals who are following in their wake. They're waiting patiently as evolution lifts humanity out of its dense emotional fog, and opens hearts and minds to the wonders they already experience. The acute sensitivity of autistics aligns them with the joy, intuition, and imagination of their high-frequency beginnings; to the magic of their creative spirit; to the radiant brilliance of their unique essence. They feel and remember their original truth; pure and loving. They believe in their innate genius, because it's palpable.

Belief is the magic ingredient that will lead you home. Your beliefs create your measure of emotion, so to untangle your emotional ties you must change your beliefs; think *differently*. Just like autistics. And this difference is reflected in the emotional centre of their brain...the limbic system. Autistics have a more highly refined and more highly evolved limbic system, simply because they're operating from a more progressed stage of evolution than their predecessors. They perceive the world from higher frequencies, having tamed the genetic emotional rollercoaster through their obstinate refusal to play the game. They see the world through the loving eyes of a higher belief system. Follow *their* lead. Step into *their* world, and watch the emotional ties of your long-held fearful beliefs fray into nothingness.

I'll be elaborating on the limbic system in Chapter Four, 'Rethinking the Brain'. However, I want to make it very clear that there's an elegant connection between evolution and autism, which is sitting quietly within the limbic system.

> The limbic system is the emotional part of the brain.
>
> Evolution is refining the emotions.
>
> ...therefore...
>
> Evolution is refining the limbic system.

Autistics are showing the world what a more highly refined and evolved limbic system looks like. They're *disabling* their emotional attachments to this world.

Primarily autism is seen to be a disability with social impairment at its core. Interestingly, upon placing the words *social* and *impairment* under the magnifying glass of etymology, a fresh understanding emerges. For social impairment boldly proclaims the *deterioration of* being a disciple to, being subservient to, being obedient to or conforming to, the dictates of others. At its core, autism champions uniqueness, creativity, individuality and difference, as it refuses to follow the crowd.
[social: see glossary]

Picasso must be delighted with autistics, as he looks down from the heavens, for they're showing the world how to preserve the genius of childhood, so it doesn't disappear in the process of growing up.

...open the door to the possibility that 'genius' might not be all that rare.

Kristen Barnett (2014) 'The Spark...a Mother's Story of Nurturing Genius and Autism.' (page 249)

Chapter Four: Rethinking the Brain

Almost everyone is born a genius and buried an idiot.

Charles Bukowski

I love this quote, primarily because it unashamedly turns the accepted norm, that 'wisdom comes with age', on its mediocre head.

Having said that, I also understand that this apparent regression into idiocy is a necessary evil in the cyclical course of evolution that's slowly but surely extricating us from the clutches of mediocrity. The forward progression of evolution sees each new generation obstinately frustrate normality, pushing its stubborn boundaries into higher knowledge and creativity, thus unwittingly reducing the depth of idiocy into which it falls by exponential increments. Yes, evolution is slowing down the fall from infancy's inherent genius into the emotional density of adulthood, within each individual. We *are* moving in the right direction, toward sustained genius. And autistics are leading the way.

With this impending onset of genius, it's imperative to rethink the brain, both in form and function. It is ludicrous to expect that the *normal* brain that serves mediocrity, could adequately facilitate genius. Through the course of evolution, something must shift within the structure of the brain, in order to accommodate this higher vision.

To fully appreciate the evolutionary shift, and its effects on the brain, particularly in relation to autism, we must first understand the term *genius* and ensure that we're all on the same page.

According to the Online Dictionary of Etymology, in which the history of words is traced, *genius* is a,

> *...tutelary or moral spirit who guides and governs an individual through life* [with its original meaning being] *generative power* [or the power that creates.]

Think *genie*. Although this appears to have a spiritual inclination, it also has a scientific counterpart within the physical body, which will become evident throughout the following pages. (you may already see the connection to the word *gene*) Once again, I've found that, in order to create a complete picture of evolution, it's necessary to address both the spiritual and the scientific, for after all, the spirit does reside within the physical body. And the physical body wouldn't exist without its animating spirit. **[gene 1: see glossary]**

It might help if you see *spirit* as *energy*. They're only words.

Genius, therefore, is the *inner* guiding spirit that powers your every creation. It's the power *you* possess to generate or create optimal life experiences for yourself. It's very attractive because it has a magnetic quality.

> *Magnetic fields sweep invisibly through the entire body, all the time.*

Maverick scientist-thinks he has discovered magnetic sixth sense humans. Science.org.

Genius is your inner tutor, whispering its inner tuition (intuition) from the highest energy field, the unified field of creation, within which we find the Earth's subtle magnetic fields. Genius fully believes in its creative powers because it knows that all possibilities already exist. There's no room for doubt on the higher frequencies because fear doesn't exist in the field of pure love, where genius resides. Belief is the magnetic essence of genius, the powerful *inner* gift, to which you must be connected in order to receive its physical manifestations. Your measure of genius, therefore, is determined by your *inner* connection to this infinite universal field of creation, your higher belief system. Genius is full self-belief, which is your own higher power.
[genius 1 & 2 & intuition: see glossary]

Am I stretching your powers of belief? I certainly hope so, for belief breathes life into the field of creation. Belief is the *energy* that powers your cells to create your physical body and your world. Belief is your spirit of genius that's tucked neatly within your genes. *You* fashion your beliefs; therefore, you are the designer of your own unique genes — designer genes!

The most distinguishing feature of genius is that it's an *inner* power. With this understanding, we can now ask the questions, what does the genius brain look like, how does the field of creation operate within the physical body, and why is autism associated with both?

Science tells us that,

> ...*the easiest way to get to know the brain is to learn the main structures of the adult brain and how they relate to its function...*

Baby's Brain Begins Now: Conception to Age 3. Urban Child Institute.

This, however, is based on the assumption that the adult brain reflects optimal functioning; that the normal developmental process into adulthood is one of refinement; that what science sees as normal is indeed progressive. But what happens if we challenge this assumption, and follow the premise that the adult brain is actually the rigid protective fortress that's been built to weather the storm of *external* stimuli. Thus providing, in plain sight, a secret hiding place for the sensitive, high-frequency infant world of genius, which cowers behind the walls of fear within each and every one of us, patiently awaiting its release through the unfolding of evolution. I do indeed believe that the adult brain is an inferior, response-based, substitute mechanism designed around fear, which only becomes necessary when you switch from your *inner* guidance system to the *external* guidance of the surrounding world during childhood. From genius to idiocy, under the guise of growing up.

By reversing the developmental process of the adult brain, and following the physiological path of fear back to its inception, we'll eventually unmask the original infant brain, with its inherent genius. As adults in pursuit of happiness, we already employ this technique to excavate our deepest fears and overcome the triggers of our childhood programming. The primitive infant stage is the desired destination.

Clearly, the genius brain looks like the original infant brain. So, commencing with the embryonic infant, let's follow the development of our clunky mechanical brain to discover exactly what causes the decline in genius, so we can subsequently reveal the evolutionary shift that's deftly engineering its progressive return. Follow closely as this will emerge like a sleeping serpent from its dark cave.

We're all just energy frequencies expressing through body matter, irrespective of the terminology that's used to describe this phenomenon. Please try to detach your personal spiritual or scientific inclinations around the language I use, and instead look for the simple meaning that I'm endeavouring to convey. Set your habitual language lenses aside, and reframe your perspective. Indeed, this is necessary if you wish to enter, and engage with, the world of genius; the world that believes anything is possible; the autistic world of higher vision, for genius animates the autistic.

As an embryonic *infant* you're connected to the *infinite* universal field of creation, which is why you radiate the highest light frequencies through your tiny bundles of tissue, maintaining their youthful complexion. Creation is always creative, never destructive. Creation begets integrity; it maintains the original design of your tissue...youth.

Infancy carries the radiant light of youth, the powerful loving guidance of self-belief and a magical imagination, all in one glorious, joyful package. In scientific terms, this is evident in the pluripotent embryonic stem cells holding the power to differentiate into unlimited possibilities. In spiritual terms, it's a powerful energetic field of grace, still free from the encumbrances of a life that's yet to come.

This was you in your infancy. You have already experienced this loving guidance; you have felt the full power of creative inspiration. You were once a shiny vehicle of light; you felt the radiant glow through your own body tissue. You were a little, joyful bundle of youth, and this experience is still imprinted within your body as your first and fondest memory. Each and every individual has this fondest memory imprinted within the fabric of their being, which is why each and every individual seeks love, joy and youth, as it calls to them from the depths of their own treasury of truth. They want to *feel* it again because they know it's beautiful; they want it to animate their physical being...as it once did.

This, too, is where you'll find the autistics. They don't want to be separated from their fondest memory, their radiant inner beauty, their truth. So, they withdraw from this dense external world because they know what's coming...the storm that buries their creative inspiration and snuffs out their light.

After months of protected bliss for the embryonic infant, a little joyful bundle of radiant genius is born. However, genius is slowly but surely programmed out of you, being usurped by childhood conditioning. Synchronously, it's during this time that the limbic brain establishes its fear/guilt, emotion-based defence mechanism, through the neuro-endocrine system introduced in Chapter Three. This is not simply a coincidence, for the limbic brain is quite literally stepping up as the protector of genius, which will be clearly explained in due course. **[limbic & defence: see glossary]**

Upon birth, the wondrous and magical land of uniqueness and possibility that's tucked neatly within the embryonic heart of every infant, finds itself in need of protection from the onslaught of external stimuli arising from well-meaning sources within the environment of the infant, intent on guiding it safely into the time-honoured boxes of growing up. The pure, high-energy state of genius that courses through the body tissue during infancy meets a world of lower energy, when delivered, head first, into the dense threads of the emotional blanket that is the electromagnetic field.

The lower vibrations of sound come crashing in, alongside grasping hands, fabrics and surfaces, all wrapped in clouds of smell, dense and foreign. A cacophony of physical and energetic laws that bear no resemblance to the light-filled creative field of the original infant-home.

Imagine the feeling of threat that's endured when this unfamiliar heaviness suddenly descends upon your safe little world of shining infancy, and follows it through childhood, accompanying every external experience. Not to be daunted, your amazingly intuitive body, having registered the danger through the highly sensitive lace-work of her delicate tissue, immediately springs into protective mode. Your *matter* is analogous to the nurturing *mother* who'll protect her infant at any cost; concurrently she's protecting her *youth*. In this external world, your matter is always swaddling your infant design (youth and creativity) and keeping it safe behind blankets of emotion, until the time comes when it can be freely expressed through your body, without threat of external reprisal. **[matter/mother, immune system 1 & 2 and immunity: see glossary]**

Thus, the limbic system is closely integrated with the immune system, the mothering system of your human body temple. Both are intent on defending your youth (your whole/healthy body) as they set up chemical borders, behind which the infant (genius, your creative spirit) takes refuge from the onslaught of external stimuli. Youth goes into hiding within the darkening cells of the human body as your inner belief system slinks into oblivion; your little light of genius is slowly being dimmed, as the shadows of idiocy loom. **[limbo: see glossary]**

Only in times of fear do you need to call upon protection and defence. Clearly, the limbic and immune systems are designed around fear; fear of the clamorous external world and all its rigid ways that cast expectation upon the unsuspecting infant. Where once the infant's body temple was a cocoon woven from light-filled threads of love, the highest creative power meeting every need through its magnetic attraction, now the need for safety and survival becomes paramount.

Throughout your life the defensive limbic system is constantly determining your level of safety, and allocating emotional values to each external stimulus that you experience, based on its bank of past memories, thus determining your chemical reactivity. The fact that it's a REaction tells you that you're *re-enacting* a past experience that occurred initially in your childhood. Your limbic brain keeps you tethered to the fears of the past, which become its memory bank. Memory and emotion are the limbic system's reason for existence.

Mmm...what do you think this means for the supposedly immature limbic brain of the autistic, who withdraws from the clutches of the stormy external world, with all its fearful memories and emotional ties? Blank stares where reactions are *expected*. Lack of *normal* social drive. Disinterest in emotional small talk. A happy present-moment demeanour. Intense focus. The *non*-limbic-controlled autistic does not bow to expectation, fear or the past. Yes, the external world is a conundrum, but they won't allow it to threaten the cherished aesthetic of their unique inner sphere. The autistic is predominantly liberating themselves from the externally-seeded fears of the past, so they've downgraded the acquired defence system of the limbic brain with all its emotional baggage. This is indeed a higher ability that would draw envy from neuro-typicals who are constantly battling with the external world and endeavouring to overcome personal emotional fears.

The heart-breaking displays of obvious emotion that are exhibited by autistics in meltdown, are the result of sheer frustration. This quote from Temple Grandin, who's probably the most well-known autistic in the world, speaks volumes.

> *I can remember the frustration of not being able to talk. I knew what I wanted to say, but I could not get the words out, so I would just scream.*
>
> Interview with Temple Grandin, *Synapse - reconnecting lives*

It begs the question: do we really need a limbic brain? The external world has a lot to answer for if our body requires a defence system to protect us from it! Clearly, *the world* needs to change...my original premise. And, thankfully, mammalian evolution is doing just that.

> *The limbic brain is acquired in the early stages of mammalian evolution.*
>
> *Interaction Design Foundation (2024), Our Three Brains – The Emotional Brain*

This mundane fact opens up a whole new can of mediocre worms.

The obvious understanding is that, in the long history of mammals and their natural evolution over time, the limbic brain made an early appearance. So, according to the macrocosmic view of evolution, the limbic brain has been around for a long time, hence its scientific name, paleo-mammalian brain, where *paleo* means *ancient*. However, with a slight shift in perspective, and with the understanding that mammalian evolution is occurring in your cells, the above quote can also be fractally applied to each and every individual human life, thusly...

> The limbic brain is acquired in the early stages in the life of a mammalian human being, that is, in childhood.

The limbic brain was NOT part of the original design. It is an *acquired* necessity in the development of an individual human being, in order to survive the external world.

Upon further investigation, we find that the word *mammal* is a cognate of the Latin word *mamma*, meaning *mother*, with a further extension to *matter*. Paleomammalian, therefore, means pertaining to the *ancient mother* or *ancient matter*, thus revealing its acquired function within the human body as the *creator of old matter…aging*.

Further to this, the word *ancient* literally means **from before**. Therefore, the ancient matter that is scribed by your paleomammalian brain is clearly the matter that's been handed down to you **from** *your ancestors who came* **before** *you*. Very simply, the paleomammalian/limbic brain reinforces your DNA within your cells through the process of childhood programming.

The family code of conduct is scribed within your limbic brain, building your *familiar* belief system and storing their habitual defensive emotions in your cells in the form of chemistry. Your belief system has suddenly shrunk to fit into the restricted family box of accepted possibilities; from genius to mediocrity. Emotions are your shrunken beliefs in chemical form; genius is in an emotional state because it's being ignored. And it's these ancestral family patterns of reactivity that deteriorate your body matter, with their fiery eruptions. DNA fosters aging. Mediocrity portends mortality. It would appear that only idiots are mortal!

So, it's in childhood that you acquire the brain that creates old matter; the brain that causes aging; the brain that angers and deteriorates your *mother*-nature, by inflaming her delicate tissue; the brain that learns to follow your family dictates and society.

Is it any wonder autistics, like Owen Suskind, withdraw from the process of growing up for fear of losing their magic.

But please don't be alarmed by this revelation, for the paleomammalian or limbic brain has an essential evolutionary purpose. It's a very courageous part of the brain, for it provides a safe haven for the refinement of the human race that's leading us back to genius; back to the universal field of creation; our first home filled with our fondest memory. The closer we get, the less refuge we require, hence the paleomammalian/limbic brain is, in effect, losing its purpose; or at least coming to the end of its protective employment and discarding its suit of emotional armour. This is evidenced in the increasing longevity that's apparent within the human race throughout evolution as the decline into aging is slowing down. The paleomammalian/limbic brain is becoming less highly engaged; resting its guard after its long defensive trek through time. **[time & karma: see glossary]**

If we look at the perceived immature development of the autistic's limbic brain, we find that it precisely reflects the less engaged, more highly evolved, paleomammalian/limbic brain model. The autistic brain more closely resembles the functions of the original infant brain, with its inner guidance system of self-belief, its creative imagination and its regenerative powers engendering youth. Mammalian evolution is deprogramming the limbic brain that creates mortality, and reinstating youth within the fabric of humanity. Mammalian evolution is snuffing out the limited family belief system, DNA, and ushering in the universal family of oneness, and the end of time.

This refinement of the human race occurs in your cells, and therefore, every single human being is fostering mammalian evolution throughout their lifespan. You're constantly contributing to the evolutionary shift; it's non-negotiable. I do hope that keeping this in mind will help you to relax your defences around change and difference, and provide an understanding lens through which you can find a softened welcoming of things that don't appear normal. For what is normal when the parameters are constantly shifting? Evolution prescribes a new *normal* in each and every new moment. **[normal: see glossary]**

The wondrous and magical land of infant *uniqueness* is the evolutionary goal. Uniqueness means *the state of being one*; not different, but all united as one on the highest frequencies of love. We're all subconsciously trying to return to this happy wonderland, within our cells. Our first and fondest memory is calling us home. Autistics just happen to be the forerunners.

Evolution is reversing the innocent misunderstanding that's been driving the human body, through fear, into mortal demise, and the so-called idiocy of adulthood. And the limbic brain is the arena wherein this transformation is evidenced, as humanity slowly awakens to the voice of the young.

Infant = *f.* Latin *in-* the opposite of + *fari* speak = **not speak**.

In parallel to its literal meaning, the infant has *no voice* during this period of childhood programming. Your spirit of genius watches on in silence. Its feelings of fear around the barrage of dense, lower vibratory external stimuli going unheeded by the adult world. The infant cannot explain that these heavy vibrations don't match its high-frequency inner guiding system, which is saying "this doesn't feel good". And those in the external world have no idea that their dense vibrations are impacting negatively upon the delicate, material, *mother*-nature of the infant. There's a distinct lack of communication simply because the infant and the external world operate on different wavelengths. An echo of the autistic/neurotypical divide.

Each new external experience triggers warning bells for the infant's highly-tuned body matter, sending a sensory message of fear to the brain; its *mother*-nature is on red alert. She's angry because there's a foreign sensation of dis-ease invading her youthful territory, and it is intent on deteriorating it. This message of fear is transmitted via electrical impulses, fizzing like hot gossip through a chain of nerve cells, imprinting their pattern as a thought that's stored within a new protein and strengthened with each revisiting; a new belief that becomes time-honoured through repetitive massaging. The firing between nerve cells as the buzz lights them up, creates a co-ignition of passion within the nervous maternal community of cells that's now primed to defend its infancy (youth, health/wholeness). The inflamed mothering system is flexing its muscles, set to fight; her immune system is now engaged as emotions run high. **[cognition, passion, belief & protein: see glossary]**

Warning bells, fear-inspired electrical impulses, fiery nerve cells, co-ignitions of passion, defensive flexing of emotional muscles and a fighting demeanour; this is your neuro-endocrine system. Surely an inferior system that provides temporary housing for the evil villain of fear, whilst banishing the innocent infant into obscurity. The feminine inclination of your pure *mother*-nature appears to have gathered a masculine edge in her wounding. A dichotomy emerges; a dual arrangement that sets feminine and wounded-feminine in opposition within the fabric of your body tissue, directly reflecting the inner and outer realms of your existence. The hidden inner feminine of your true infant nature and the publicly wounded outer feminine (masculine) have taken up residence within each and every cell of your body. Your loving and gentle infant nature is masked by your fearful and angry defensive limbic nature. Duality has replaced unity. **[adult & duality: see glossary]**

Throughout childhood, as you're being programmed by the traditions of family, your wise and wary body matter meticulously registers the unique patterns of each external experience you encounter, creating an ever-increasing bank of thoughts that's set within your strengthening protein structure. This bank of memories becomes your personal record of beliefs that sets the standard of accepted familial expectation - the family code of conduct – guiding all your subsequent responses to the world around you, and dispatching appropriate chemical soldiers that settle in your cells as emotion. Proteins (habits) power all your chemical reactions. Your limbic system is born, along with your individual life story, your armour of *normal*, the illusory tale that becomes your reality through its perpetual retelling.

But this story is a disguise that's authored by the family, and is founded on fear. The family are telling tales in order to maintain tradition, for they believe that tradition is safe; familiarity is safe. Family tradition beds down within your protein structure; your body matter must be obedient to the family patterns; hence, similarities in family physiology. Your individual life fairy tale, as scribed by your limbic brain, is your protective and defensive mask of fear and emotion, otherwise known as your persona, which keeps you in servitude to the external world. You've learned to read the signals and accommodate the rules, staying safely within traditional *boundaries* of expectation. Isn't it interesting that the word *limbic* literally means border, which also means boundary. The limbic brain is the storehouse of normal; your limited, mediocre world. The family way of life, as you know it, creates normality and mediocrity out of habit, the arch nemesis of creativity and genius. The family way of life is indeed a mask that's concealing your true identity.

The Oxford Dictionary of Etymology tells us that the word *persona* means *a mask used by a player, a false face*, stemming from the word *person*, which means *character, part played; human being*. **[persona: see glossary]**

> *It's formed as an adaptive, learned, defense to the true self.*
>
> *Lancer, Darlene, JD, LMFT. The Co-dependent False Self*

As your infant nature slips behind the mask of your limbic-inspired persona, you slip on the false face of your human character in this theatre of life, playing your part as per the family script (DNA). Your role is defined by the set of specific characteristics, beliefs and behaviours that were honed during childhood. The magical and magnetic universal field of creation is replaced by the electro-magnetic field of emotion, as the vast offering of possibilities dwindles to a limited set of personal family favourites. Fear and limitation are your new guides in your role as adult. Masking is a necessity for autistics to survive in this neurotypical world. Can you see that the masks chosen by autistics to navigate each situation must correspond to the specific personas of the neurotypicals with whom they're interacting? Autistics mask *because* neurotypicals are masked. It takes a lot of energy to pretend to be what you're not; to hide your true essence; your infant (infinite) nature.

The infant has no limits; the world teaches them limits. The infant has no fear; the world teaches them fear. The world fashions their persona, their mask, their illusion. The illusory world of fear imprints its limbic mask firmly upon your true infant design, and your reptilian brain emerges, fiery and instinctual. **[illusion, evolution & reptilian: see glossary]**

Infant Brain + Limbic Brain = Reptilian Brain

Infancy, wearing her limbic mask of defence, portrays the wounded persona that's expressed by your reptilian brain; your individual personality with all its fears and emotions, passion and anger. As she steps into adulthood, the infant finds a voice, albeit an emotional voice that's wounded from the trauma of childhood.

The infant brain wearing its false limbic mask of repetition and instinctual behaviour, personifies the mythological fiery dragon protecting its treasure within the dark cave of your cells. All your chemical reactivity is reptilian, including competitiveness, sexuality, anger and defence. Clearly, your inflamed mothering/immune system, who's primed by passion as she protects her treasured infant, wears the cloak of the ancient beast; the reptilian.

The fiery reptilian mother is the defensive hormonal blanket of fear and emotion that swaddles the infant within your cells, and unfortunately, also serves to deteriorate its matter through chemical reactivity. The paleo-mammalian or limbic brain marks its presence, as your hormones mask your truth.

The highly sensitive *mother*-nature (matter) of the autistic, no longer wishes to be sacrificed by blindly following the dictates of the external world. Autistics don't want to embody the fiery reptilian; they're far too gentle to carry that dense and demanding load. Only when disrespected by an ignorant mediocrity do they exhibit reptilian-like characteristics.

Autistics simply want to align with their truth; they feel that truth within every fibre of their being; the truth that radiates from the highest light frequencies; the truth that is their first and fondest memory; the truth that is the evolutionary finish line; their infant/infinite *mother*-nature in full expression.

The eminent neuroscientist, Dr Paul D MacLean, who is famous for his triune brain model of evolution, said that...

> *... the idea of the limbic system leads to a recognition that its presence "represents the history of mammals, and their distinctive family way of life".*
>
> *Limbic System, Wikipedia*

Although science has advanced its understanding of the triune brain model, the functions of the limbic, reptilian, and neo-cortex remain current.

The history of mammals, and their distinct family way of life, is represented in the limbic brain.

Family dictates the *familiar* or habitual patterns of belief and behaviour; the repetitive limbic patterns of fear and emotion that articulate through the worldly presence of your reptilian nature, your persona. Family is reptilian. The limbic system, therefore, records the evolution of the fiery reptilian, as she creeps from generation to generation within the fabric of your *mother*-nature, your matter. Fortunately, her fury is being tamed with the progressive generational awakening of love within the family way of life, and indeed, the world. Soothing strains of this higher belief system increasingly relax her maternal defences, as she gently releases her long-held youthful treasure from its place of hiding within the dark cave of your cells. Mother and infant are reuniting; matter and youth embracing.

This is evidenced through the increased longevity of the human body throughout evolution as the increasing flow of higher vibrations within the cells slows down the aging process. The long evolutionary trek, that's been scribed by the limbic system within the threads of the human body, also epitomises the perpetual search for the fountain of youth, which can only be attained from within. It's reinstating a higher belief system with your reconnection to the universal field of creation, your inner source of higher power, your self-belief. Evolution is liberating the integrity of mother nature by empowering her within each and every individual.

Evolution is taming the reptilian, the family, and disarming the defence system. Humanity is emerging from the dark cellular cave of fear, throwing off its impassioned limbic mask of protection and revealing its true design. As evolution's forerunners, autistics are helping to reinstate a world without fear, which means a world without the limbic brain. They're not introducing a new system, just refusing to follow the family into the clunky mechanical adult brain; refusing to leave their original infant design and lose their magic; choosing to hold onto their genius, their light. Autistics are showing us how to sustain genius in a world of mediocrity. Autistics are teaching the world to raise their vibrations into love, and embrace authenticity. And when the world around us is loving, we don't need a defence system for protection. We don't need a limbic system.

The protective duties of the limbic brain throughout evolution have provided us with a personal store of familiar beliefs, and a family haven within which we could safely overcome our emotional proclivities.

Humanity has slowly but surely been rising out of the illusion of dense vibrations that manifests in an unhappy world of mediocrity. Higher vibrations have opened us to higher information, bringing progressively inspired scientific and technological advancements raining vast improvements upon us. Healthier food and lifestyle choices, reduced noise density, better sleep options and improved housing, have all ushered the family way of life into greater ease and enjoyment. Better understanding of the human body, increased nurturing of the human spirit, greater acceptance of differences in culture and religion and more refined skills in the area of creative arts. All these advancements, both spiritual and scientific, have contributed to the soothing of your *mother*-nature, and the refinement of the limbic brain with its historical account of the family way of life through the generations.

As everything external is evolving, it stands to reason that the limbic brain responds accordingly. The courageous limbic brain, which has written the fairy tale of evolution upon the parchment of the human body, is now striding towards its happily ever after, one individual at a time. **[information: see glossary]**

In following the development of our clunky mechanical brain, we've discovered that *the world*, primarily the family, causes the decline in genius that transpires from infancy into adulthood. It's the switch from inner loving guidance to the external guidance of the family and society that causes the fall from the high-powered vision of genius into the limited, approval-seeking world of mediocrity that we call normal. **[mediocrity: see glossary]**

The shift that's deftly engineering the progressive return of creative genius is the *inward* shift that's leading us back into the loving nature of our true essence, the higher vision that's already indelibly woven within the very fabric of our being. The inward evolutionary shift is a shift toward self-belief, as we remember with fondness our true design and give ourselves a little hug of self-love. I remember you.

In terms of physiology, the shift occurs in the limbic brain. The limbic brain is the only place you change your story. When you change your pattern of response to an external situation, you over-ride your old memory; you change what you believe about that situation. By changing your stored memory, you change the energy in your cells, as well as your gene expression and protein structure. You literally change your beliefs about who you are in relation to your environment. You instate a new familiar, and evolve the family way of life. Evolution transpires in your limbic brain, and expresses through your cells. Your physical body is the vehicle of evolution.

I stated in Chapter Three that there is an elegant connection between evolution and autism, which sits quietly within the limbic system.

> The limbic system is the emotional part of the brain.
>
> Evolution is refining the emotions.
>
> ...therefore...

Evolution is refining the limbic system.

The emotional memory system of the limbic brain is shifting, literally. It's stepping aside and making way for a new family way of life; a universal family where everyone is united. Autistics already unknowingly embrace this one big happy family model, which is the very thing that sets them apart from the norm. The limbic brain of autistics is simply more highly evolved.

Evolution is the incredibly well-designed individual pursuit that's guided by personal emotions and beliefs. We're all evolving at different rates, which is reflected in our brain structure, and accordingly in the structure of our physical body. Common sense would indicate that the brain structure of each individual human being is uniquely molded by their particular personality. Unfortunately, mediocrity has hypnotised us into believing that there is such a thing as a *normal* brain structure that we must all exhibit similarly. No nuances; no room for individual differences. *My* brain tells me this is not so.

Does the average brain of the twenty first century look identical to the brain from five hundred, a thousand, even two thousand years ago? I doubt it. But how do we know, as the technology of the past was not capable of providing such evidence. With improved technology, science can now brilliantly present us with the minute biological details of our brain. However, with no means of comparison, it's assumed that the current brain structure is normal, with any variation being faulty, disabled. This is the fate of autistics.

Enough! It's time to rethink the brain, from a higher perspective.

Genius is simply hiding behind the mask of the normal brain; Sleeping Beauty awaiting the return of her magnetic super power, the kiss of self-belief that rides the highest frequencies. It's your inner beauty that's sleeping, and your own princely charm of self-belief that breaks the spell.

Sleeping Beauty is awakened in the brilliance of autistics. They don't doubt their ingenuity. Only the world does that. But why would anyone deny the obvious? Perhaps because it threatens their antiquated belief system?

Fear not, for the evolutionary shift is bestowing genius upon *everyone*, not just a selected few. When you step out of your own fear-filled way, and welcome a new perception that embodies self-belief, your super powers emerge, just like the fiery dragon from its dark cave and Sleeping Beauty from her enforced slumber.

Are you ready to peek behind your mask and embrace your own creative genius, by believing in yourself?

You're an idiot if you don't!

> *To be yourself in a world that is constantly trying to make you something else, is the greatest accomplishment.*
>
> *Ralph Waldo Emerson*

Chapter Five: A Surprising Twist

Look deep into nature, and then you will understand everything better.

Albert Einstein

A surprising twist is the mark of a good story; it keeps you on your toes with its sharp choreographic swerve. You weren't expecting that! Just as you weren't expecting autism to be the frontrunner in the story of evolution.

Chapter Four revealed that evolution is set neatly within the limbic brain, where memories and emotions dance to the tune of family beliefs, thus determining your physiology and your reality.

Clearly, your spiritual proclivities perform an intimate pas de deux with your body matter; a delicate ballet of spirit and science.

However, the very idea of spirit and science embracing, can cause unrest. For science has been reluctant to dance publicly with spirit in the past, relegating it to the realms of fantasy; a repugnant wallflower that's not fit to participate in the dance of life. But this is exactly why the autistic pas de deux is off kilter; it's out of balance; wobbling through time like a broken-down music box ballerina.

It's true that science tells us there's something different about the physiology of the autistic brain, but it's spirit that explains why (due to the evolution of family belief patterns – the pioneering spirit of innovation that sits within us all). Without the *why*, autism has been taken out of context, and therefore makes little sense. And science is bewildered.

Science without spirit is like a one-handed clap; it's incomplete. The spirit/science conundrum is clearly evident and accepted within quantum mechanics in the form of the wave-particle duality, which states that all objects have both a wave *and* particle nature. With an open mind and just a little dash of common sense, an analogy can be drawn wherein the *wave nature* (energy) aligns with the concept of *spirit*, and the *particle nature* (matter) aligns with the concept of *science*. All objects, therefore, possess a spirit/science duality; where there's one, there's the other; they're inseparable, which makes it impossible to leave either one out of the equation when seeking the truth about reality.

Your physical body is a perfect example of this wave/particle duality, where your body matter takes its form from your energy or beliefs. Your body is, in fact, the *scientific* evidence of your *spiritual* inclinations. And accordingly, if you change your spiritual inclinations, you change your body matter. This is precisely how evolution transpires!

With this understanding, what becomes apparent is that science follows spirit. Just as waves collapse to form particles, your energy dictates the *particular* form of your body matter. Beliefs create reality. Spirit, pioneering or otherwise, sets the ground upon which scientific fact is founded.

Therefore, science will never find the answers to the autistic conundrum until it looks in a spiritual direction, and recognises the impact that beliefs have upon the brain and consequently the body. Then perhaps, after lavishing focused and careful attention upon the unique beliefs of autistics themselves, scientists might discover that autistic brains function as they do because they're big-hearted, out-of-the-box thinkers, who believe in a united universe.

Thankfully, neuroscience is beginning to provide the much-sought-after evidence to show that your beliefs, indeed your energetic or spiritual nature, govern your genes and therefore the protein structure that makes up your body matter. The form of your body protein is designed by your religiously practised beliefs, courtesy of your genes. Your proteins are simply habits inscribed upon the parchment of your body tissue. So, what habits have evolved to prescribe autism?

> *Typically, autism cannot be traced to a Mendelian (single-gene) mutation or to single chromosome abnormalities...*
>
> Heritability of Autism, Wikipedia

It's pointless to seek a universal gene that explains the habit called autism, when the expression of genes is influenced by *family* beliefs, simply because all family's beliefs differ. This does, however, explain why autism expresses so differently from family to family; from individual to individual. A gene, which is considered to be the basic physical and functional unit of heredity, is actually analogous to a genie in a bottle, carrying out your every wish. The gene follows your beliefs; science following spirit. Yes, genes pass on the family data from generation to generation, but only because of the religiously practised beliefs and habitual behaviour patterns that have been programmed into your energetic system during childhood, and continue as your spiritual inclination, until *you* deem otherwise; until *you* instruct your inner genie differently. And isn't this exactly what happens within every new generation. As you generate new beliefs, a new generation is born. Autism is simply the next stage in the evolution of beliefs, inspiring an inner genie that's more magical and more powerful than in previous generations, which is why it's misunderstood. **[generation 1 & 2: see glossary]**

I've recently discovered the scientific concept of *memes*, in relation to the understanding of autism. These are not to be confused with the humorous images and videos that spread virally on the internet, although they follow the same pattern of infectious behaviour. A scientific meme is considered to be a 'unit of cultural information', such as a concept, idea, belief, behaviour, style, fashion or practice, which spreads by means of imitation from person to person. The term was coined by Richard Dawkins in his book The Selfish Gene.

> *We need a name for the new replicator, a noun that conveys the idea of a unit of cultural transmission, or a unit of imitation. 'Mimeme' comes from a suitable Greek root, but I want a monosyllable that sounds a bit like 'gene'. I hope my classicist friends will forgive me if I abbreviate mimeme to meme. If it is any consolation, it could alternatively be thought of as being related to 'memory', or to the French word même. It should be pronounced to rhyme with 'cream'.*
>
> *Richard Dawkins (1976) The Selfish Gene*

Effectively memes are what we massage into being; what we nurture, improve upon and refine; what defines individuals, families, communities and countries; what's stored in our memory banks; our limbic records. Mammalian evolution is inherent in the meme. Autism embodies ahead-of-the-curve memes.

Quite frankly, I believe it's a thinly disguised attempt to employ a scientific label for anything that comes under a spiritual umbrella; that which defines the essence of a human being. And even though I've never heard the term before, it would seem that this book is inadvertently filled with memes, through which I unashamedly illustrate autism in relation to evolution. I'm not afraid to embrace the spiritual aspects of our nature, and call them what they are, because without them we wouldn't exist; and because it simply makes sense. I don't feel the need to re-label them in order to give them a scientific identity. Science still can't openly align itself with the spiritual, but at least it's now recognising its presence, albeit covertly tucked behind the *meme* mask.

Science deems that autism is explained more completely by memes than by genes. This would infer, by my calculations, that autism is explained more fully by the spiritual than the scientific; that autism is explained more rightly by its essential, rather than its physical, nature. Something I'm certain is an unintentional bequest of the meme creation!

I feel it would be more apt to say that the spirit/science connection is analogous to the meme/gene relationship, with memes representing the spiritual, and genes representing the physical or scientific manifestation through which the meme is expressed. The meme is the animating expression of the gene, and therefore *everyone* is explained more completely by memes than genes. The evolution of culture is expressed through the human body, with autistics displaying more highly evolved memes; a new culture that's as yet misunderstood. It stands to reason that both memes *and* genes explain the presence of autism, and this understanding in relation to evolution is paramount.

So, science fixing autism is a misnomer, for there's nothing to fix, except a very big misunderstanding. Science can, however, fix its own perception of autism by discarding the spiritual blinkers.

In order to do this, it needs to look deeply into nature, as suggested by Mr. Einstein in the quote at the beginning of this chapter.

So, let's start with the following list of characteristics, which I'm sure you'll recognise:

*An acute sensitivity to sound, smell and touch.

*A deeper inner state of consciousness.

*A feeling of disconnection from the 'real' (mainstream) world.

*Savant tendencies...higher states of consciousness and knowledge beyond that which has been taught.

*Feel the negative energies of people and situations more acutely, and consequently get more easily upset.

*Fussy about what foods they eat.

*Clear visions...thinking in pictures.

*Extra sensory hearing ability.

*Intellectual curiosity, with intense interest.

*Increased intuition e.g. correctly naming musical notes by tuning into their frequencies.

*Introverted personality.

*Very kind and forgiving.

*Truth is paramount. Sense true intentions of others energetically. Very connected with nature...all living things e.g. animals.

*See beauty in the patterns of life.

*Heightened concentration and focus.

*Attention to detail...precision.

*Non-verbal communication...akin to telepathy via tuning into vibes.

*Self-directed and taught.

*Heightened and prolonged delta brainwave patterns.

This is a list of autistic characteristics...right?

Well, here's the surprising twist...hold onto your hat!

They're actually the characteristics attributed to someone who has *opened their third eye*, a supposedly spiritual phenomenon. The opening of the third eye and autism bear a remarkable resemblance, yes? Autism is forcing science to take a deeper look at the nature of the human body, where *taking a deeper look* involves holding a magnifying glass over its hitherto neglected spiritual landscape. For those with a scientific inclination it might help if you view the opening of the third eye as simply being a term that spiritual communities use for the purposes of communication. The fact that it's the *spiritual* communities that are aware of this higher vision, doesn't make it any less physical.

In fact, I'd like to introduce the understanding that the autistic brain provides the scientific evidence for the opening of the spiritual third eye.

Far from being just woo-woo, the opening of the third eye actually registers within the physiology of the human body, and has been doing so progressively throughout evolution. When looking through the spiritual magnifying glass to reveal this physical evidence, we journey through the brain's third ventricle, and find ourselves back in the limbic brain, the domain of autistic difference, with a particular focus on its relationship with the pineal gland, which will be discussed in detail a little later.

I personally believe that the brain's *third ventricle*, around which the limbic village nestles, is the hub of the *third eye*. **[window, third eye & third ventricle: see glossary]**

It would appear that the third eye is intimately linked to both autistic difference and the evolution of family belief patterns (DNA), due to its connection with the limbic system. Therefore, the scientific concepts of autism and evolution have a direct relationship with the spiritual concept known as the opening of the third eye.

The word *open* literally means to *un-pen;* in other words, to release something from an enclosure; to set it free; so that it's no longer hidden, secret or mysterious, in the closet or fenced in.
[open: see glossary]

So, the opening of the third eye represents the liberation of the third ventricle from the clutches of its fear-inspired, defensive limbic borders.

The opening of the third eye aligns with the *un-penning*, or releasing, of your stored defence records from within the memory banks of the fortified, protective stronghold of your limbic system. Evolution has been progressively updating your family records of fear (DNA) in an ongoing cellular cleansing. Your limbic memory banks are being wiped clean, along with all their prescribed limitations, rules, restrictions and genetic codes of conduct; dis-ease, emotions and struggles. Evolution is a long-running epic of forgiveness, where the memories of the past are cleansed from within the body tissue of humanity. Evolution is a cellular baptism.

The autistic brain exemplifies the advanced model of forgiveness that's associated with the opening of the third eye. By opening the third eye, evolution has created the enlightened state called autism.

Before you get too hung up on the word *enlightened*, let's simplify this forgiveness process in relation to your physical body as it journeys through evolution.

Bear in mind that the word forgiveness simply means letting go of, or completely giving away or releasing. In this case, the object of forgiveness that's being cleansed from the cells of the body tissue, is fear.

1. The Nervous System is relaying less fearful messages in response to the environment, as outdated family patterns of belief are overcome.

2. Reduced fear necessitates less defences, so limbic activity progressively reduces.

3. Less chemical soldiers are dispatched by the endocrine system due to reduced fear and lower defensive activity. Body chemistry is refining, purifying, along with the essence within the cells of humanity.

4. DNA in the cells is evolved through these epigenetic modifications in your gene expression, where dis-ease is forgiven and your body tissue is relaxed into a state of increased ease, otherwise known as youth.

Evolution is soothing the nervous system, disarming the defensive limbic system, deactivating the chemically reactive endocrine system and cleansing the cells of ancestral genetic threads of dis-ease.

Evolution is rendering fear, defensiveness and chemical reactivity obsolete, as it transports the family code of conduct (DNA) through time. So, DNA is being transported through time within the neuro-endocrine and limbic systems, on a mission of cleansing your cells of inherited family disease. The family *dis-ease* is being forgiven, released, let go. The family is undergoing a genetic baptism. DNA is being refined within the electro-chemical system of your body, through the evolution of your cultural dictates, your spiritual proclivities, your memes.

> *...memes are random electrochemical activity in the brain...*

Henry Kong (2006) More Self Than Self: At Autism's Edge (page 203)

With a slight shift in perspective, it becomes clear that there is actually no *linear* construct called evolution, but that DNA is literally evolving uniquely within the neuro-endocrine and limbic systems of each individual. Your one precious lifetime gives you the opportunity to contribute to this evolutionary cleansing of fear by releasing the emotional prisoners that have been clogging up your cells; by switching your focus from emotions to feelings.

Evolution, through the opening of the third eye, is cleansing your body tissue, thus creating an increasingly pure canvas for the reception of progressively higher light frequencies. The autistic canvas is the case in point, for they have indeed chosen to tune into their present-moment feelings rather than the emotions from the past.

The frequencies that your body is *feeling* right now, give you the true information about the present situation. You just have to stop and become aware of them. Then by choosing responses that honour those feelings, you nurture both your body and the world around you. You've listened to what your body is telling you, and she shows her appreciation by feeling good and glowing with increased youth. Instead of reacting emotionally from the neuro-chemical dictates of the past that are lingering in your memory banks, you always have the choice, in each moment, to invite feel-good vibes into the situation; to change your perspective and evolve your behaviour and your physical structure. Paying attention to your feelings is the key, and your senses are your guides. **[nurture: see glossary]**

When you soothe your fearful, nervous inner voices, you render them inert, or still. This is how the nervous system is being soothed and quieted through evolution. Every time you soothe a fear, you short-circuit the automatic cycle that perpetuates your chemical reactivity. By doing so, you soothe your body tissue with the gentle higher frequencies that help to maintain its integrity; you employ the energies of creation rather than destruction. Nurturing your body tissue fosters youth within its structure and joy within its spirit. **[joy & fruit: see glossary]**

You can be the creator of your life by nurturing your feelings. Or you can remain a spectator by succumbing to your emotions. Creators contribute to evolution; spectators stay the same. The inward evolutionary shift is making creators of us all, as we progressively move from externally driven automatic reactivity to a more conscious internal choice. From emotions to feelings. That's the goal of evolution. To clear your cells of emotion so you can tune into your true and immediate feelings and savour the gentle touch of those loving high vibes.

Autistics, with their acutely sensitive body tissue, are living proof of this incredibly ingenious plan. Autistics are guided by their feelings, not by cultural and social conformity to the external world with all its emotional attachments. Autistics are creators, not spectators.

But they're ahead of the time, quite literally, and therefore must carry the heavy burden of all trailblazers. They're putting their acutely sensitive bodies on the line, as evolution's guinea pigs, until the world breaks through its long-standing genetic walls of fear. They're suffering due to the ignorance of those who do not understand the plan of evolution. Strangely enough, even *that's* part of the plan! Without suffering, nobody notices. But when children are suffering, hearts are broken, and action is taken. The precious, but difficult, lifetime of each autistic is a heroic contribution toward the realisation of evolution's plan, and the truth about the concept we call time. For all is not as it seems.

Your own individual lifetime is all that exists of time...for you. We're all hypnotised into believing that time is a continuum that began eons ago, and that it moves constantly into a far distant future. However, I believe that's simply an illusion. For time

begins *again* with the conception of each and every individual human unit; is played out through electromagnetic light frequencies within the temple of the body; and ends with the depletion of the individual's energy. You're operating like a battery, where *the-story-of-this-lifetime* is your energy. Who you believe yourself to be, is the energy that powers your body and creates your reality. Your lifetime experiences are stored as emotion in the memory banks of your limbic system, scribing your life-story into an electromagnetic motion picture of light...your reality. Throughout evolution, the limbic script has been perpetually reimagined, its storyline reflecting the magnified creative vision of increased beauty and wonder within each successive generation. As the area of difference within autistics, this understanding matters. Let's take a closer look.
[emotion & time: see glossary]

[In this instance, I feel it's necessary to include the word-sequence for *time* in the chapter as well as in the glossary, as I believe it will help if you refer to it during the following explanation.]

TIME = TIE M = TIE EM = TIE ELECTRO MAGNETISM = E‿M

I'll now endeavour to explain TIME as clearly as possible, because it shows the crucial link between autism and evolution.

In the word-sequence above, the **limbic system** is represented by the little **tie** or slur that links the **E** and **M** at the end of the sequence. What's most important in understanding this concept is that the limbic system *links* the neuro-endocrine systems within the brain. The neuro, or nervous system, is the body's *electrical* system (**E**), and the endocrine, or chemical system, is the body's *magnetic* system (**M**). Therefore, in linking the neuro-endocrine systems, the limbic system *ties* your body's *electro-magnetic* systems together. **TIE EM** or **E‿M**

In so doing, the limbic system records TIME within your cells in two distinct ways.

Firstly, it records, within the memory banks of the individual, the mammalian progression through evolution, marked by the successive changes in the family's code of conduct (DNA) through the generations = **(a) time** marching into cultural modernity through changing belief patterns.

Secondly, it determines the physical deterioration called aging = **(b) time** imprinting itself upon your body tissue.

Although **(a)** and **(b)** are distinct, they work together, as your family story energises your behaviours and shapes your body tissue. Your history determines your measure of aging and therefore the condition of your body tissue. The family beliefs that have travelled through *the ages*, dictate your individual patterns of *aging*. Your body ages in direct proportion with your beliefs (and accordingly with your level of *self*-belief).

So, the limbic system is where your electro-chemical batteries (cells) charge and deplete; where *the-story-of-this-lifetime* animates your body according to your beliefs; where your lifetime is measured in units of energy-sapping fear and emotion. **TIME** sits within your limbic system.

It appears that the limbic system is responsible for both time and space.

Increased longevity throughout evolution occurs within the memory banks of your limbic system through the progressive deletion of its time-worn fearful memories. MEMORY & EMOTION are functions of TIME. Both feed on fear. So, time, memory and emotion only exist because of fear.

> *Our senses are the port of entry for all memory, stimulated by our external environment.*

Caroline de Braganza, Revealing the Truth about our Memory Banks

The records of our five senses are stored in the limbic system. The memory function of the limbic system exists solely in *response* to the external environment. Therefore, TIME too is an internal bodily construct that's created by *your reactions* to the external environment. You create your own lifetime by virtue of your emotional reactions to external stimuli, which become the story of your life; your store of memories. TIME is a sensorial fairy tale of *your* creation.

Take away your habitual reactions to the impacting external environment, and what do you have? Something that looks very like autism. The limbic system disappears, along with the sad electro-magnetic stories that were once perpetuated in your deteriorating neuro-endocrine systems.

In its stead, you have the *inward* evolutionary shift, which is crafting a new story within the cells of every individual; a story that's based on increased *self*-guidance and nurturing, instead of external expectations and control. As you dive within, and listen to your nurturing inner voice, you find a wellspring of inspiration; your triumphant inner spirit, rising like a phoenix, from the ashes to ashes, dust to dust electro-magnetic cycles of time, memory and emotion. You are opening to your sixth sense.
[phoenix & embryo: see glossary]

As you dive within, you discover that the antiquated memory function of your clunky mechanical brain serves only to remind you of what *others* expect, so you can emulate and regurgitate; and that this inferior intelligence system is becoming defunct. When you stop reacting to the external environment, and quietly tend your internal garden, honouring its desire to feel good, you invite the pure magnetic field of creation into the fabric of your being; you open to a world of creativity beyond your wildest dreams; you *un-pen* the unique imaginings of your inner spirit; you liberate an enlightened vision and project it upon the physical canvas, creating a new world reality. The spiritual well of creative genius is instantaneous and eternal, when you ride the highest light frequencies of self-belief. This is not the neurotypical world with its limited possibilities, rigid rules and social strictures. This is a world far beyond the reaches of mediocrity. A world that's accessible only through the portal of your imagination...evolution's inward destiny. Evolution is shifting your source of power from the electro-magnetic clunky mechanical brain to the fully magnetic imagination, within each individual. **[magnet: see glossary]**

Neurotypicals create their life story based on external experiences, fashioning a shallow limbic well of emotion into which they repeatedly dip to be guided by the past. Autistics create their story from the deep inner font of their infinite imagination. Their inner spirit of creation is their guide. The *story-of-this-lifetime*, or energy, of autistics, is far more expansive and imaginative, bestowing upon them the many savant gifts that are considered super powers, but which are actually available to all human beings who are ready to throw off their limitations and believe in themselves. **[infant & infancy: see glossary]**

Isn't this inward shift the very accomplishment that's been sought after and relished by all spiritual advocates throughout history? And isn't this connection to your inner creative spirit exactly what everyone seeks in their pursuit of happiness? Then why is it so hard to recognise and accept this shift within autistics? Autistics are introspective, self-motivated, self-directed, inwardly focused creatives. Autistics show both the spiritual and physical signs of the evolutionary shift into genius and love; they're connected to their inner well of creative inspiration, their limbic systems display decreased activity and they thrive in their happy little world of no-time. And yet they're considered disabled. It's time to mend the wobbly autistic pas de deux. It's time for science to fully embrace its spiritual partner, and indeed shift its own broken beliefs. Perhaps it's time for science to start believing that there's more to autistics than what meets their blinkered eyes. The world is watching you, science, and following your lead. The world believes what you tell them. This is your opportunity to discard the blinkers, to open your eyes and embrace the truth about autism, so the world can embrace it too. When science embraces spirit, wonders will never cease...literally.

Autism, through the opening of the third eye, prompting the discarding of antiquated limbic borders, is indeed liberating the hearts and minds of humanity and creating a higher collective vision and a new world reality. A world in which all individuals are accepted and accepting; where difference doesn't exist; where self-belief reigns; a world where each individual is free to express their true essence without fear of retribution or disdain. As I've already suggested, this heart-centred world reality is inherently embodied by autistics with their kind, nature-loving spirits, clear vision, wonder and curiosity, acute sensitivity and deeply profound sense of spiritual integrity. Autistics see and feel the beauty in the pattern of evolution. **[beauty: see glossary]**

What science calls *evolution*, the spiritual world calls *ascension*. Same, same. Just different words used to describe the same phenomenon. Ultimately, they're both about rising beyond, or coming-up-over (overcoming) difficulties, into higher vibrations, visions, power, light, energies, beliefs...realities. As humanity progressively *evolves* its beliefs into a higher vision for itself, its vibrations *ascend* into the dizzy heights of creative genius, with all its accompanying characteristics. The world is lifting the ancestral veil of illusion that's had us believing in our fears, and gently revealing the light of truth that is our happy destiny. The opening of the third eye is occurring without your conscious awareness; it's simply inherent in the natural course of evolution, autistics being its finest example. **[evolution: see** glossary]

And yes, this is going to require the suspension of your disbelief. However, isn't that the very nature of evolution? Doesn't evolution require the pushing of boundaries into previously unexplored territory, which often *seems* unbelievable…at first?!

What seems unbelievable to me, is the fact that evolution has been occurring for soooooooo long, and yet we humans still resist it at every turn. Haven't we learned the pattern yet? Someone takes a little sojourn into the imagination and theorises an *unbelievable* idea. We humans, who are creatures of habit and therefore prefer our comfort zones, resist tooth and nail, saying "How can this be possible?" After much testing by the imaginative pioneering spirit, the idea is finally proven. *Now* it's called science. And *now* we humans believe it, because the scientists tell us it's true. But it was *always* true. All that resisting! What a waste of energy! I'd have thought by now that the most obvious response to a new idea would be "Anything's possible…let's find out".

On that note, let's find out about the pineal gland and its relationship to the third eye, and indeed autism, as promised earlier. The twist continues in the next chapter, with an intriguing connection between the pineal and Pinocchio.

Chapter Six: The Pineal and Pinocchio

Secret longings underpin the beautiful fairy tale Pinocchio, wherein Geppetto's deeply heartfelt wishes inspire his wooden puppet to come to life.

Secret longings are also the auspices of the tiny pinecone-shaped endocrine gland called the pineal, which has an intimate association with both autism and evolution...providing a fascinating, and unexpected, connection with Pinocchio.

In reality, the pineal sits boldly unpaired along the midline of the brain at the posterior of the third ventricle, from where it is wired to the limbic system, the area of difference in autistics. Already we can see that the pineal is connected to the opening of the third eye *and* autism, by virtue of its location and connections. **[pineal 2 - 6: see glossary]**

According to neuroscientists, the pineal is a neuro-endocrine transducer, which simply means that it converts one energy form into another. In this case, the pineals most obvious duty is to convert the electrical energy of your nervous system into the chemical energy of your endocrine system. Your inherent familiar belief patterns dictate the electrical firings of your fear-based nervous system, sending messages of alarm to your pineal, which accordingly deploys the appropriate chemical soldiers to your organs, prompting your emotional responses. The pineal appears to be a major contributor to the fear-guilt cycle of your neuro-endocrine system. But the pineal has no power of veto. It's simply doing what it's designed to do. The pineal processes incoming information and automatically sends out chemicals based on your stored records from the past. Fear creates memory, and triggers emotion. The pineal can be likened to a chemist, who dispenses chemical concoctions that match the incoming script. The chemical concoctions are designed to fight the offending dis-ease that's posing a threat to your body's equilibrium based on your stored memory of past experiences. Your pineal is integral to creating *the-story-of-this-lifetime* that projects your electromagnetic motion picture of light into reality, as discussed in Chapter Five.

However, this is the pineal operating from its sleepy state, where it's informed by fear...the fears of the past that are recorded as your limbic memory. Evolution is slowly but surely liberating the dormant pineal and reinstating its optimal functioning. The fully restored pineal is a transducer that converts the highest light frequencies into vibrant images, just like a television, only inside your head. This is often called the mind's eye, which has a direct correlation to the opening of the third eye and to autism. I'm immediately reminded of the book called Thinking in Pictures by well-renowned autistic, Temple Grandin.

> *I think in pictures...I translate both spoken and written words into full-color movies, complete with sound, which run like a VCR tape in my head.*

Temple Grandin (1995), Thinking in Pictures (Chapter 1)

These apparent flashes of brilliance that dance in the minds of autistics like Temple Grandin, are the work of the third eye, and indeed the pineal, which has long been associated with altered states like dreaming, religious visions, near-death experiences, telepathy and the like.

These phenomena occur when you connect with the higher frequencies of your unconscious mind, which exists beyond time and space; beyond the restrictions of your limbic memory banks; in the field of infinite possibilities. The open third eye is a well of creative solutions, where the limitations of your waking reality don't exist; where the pineal is no longer fed the illusion of fear through limbic connections.

The opening of the third eye expresses uniquely within individuals, displaying a wide variety of characteristic gifts, each of which has been given an appropriate label, which I'm sure you'll recognise.

Clairvoyance is the most obvious example, where information is transferred through extraordinary *visual* experiences, as with Temple Grandin. Other examples include clairaudience, where information is received through paranormal *hearing*; claircognisance, through spontaneous *knowing*; clairsentience, with its extraordinarily refined sense of *feeling*; or any combination of these. All are marked by the same heightening of the senses that's demonstrated so naturally within autistics. Acute sensitivities in sight, hearing, smelling and feeling, extraordinary skills in mathematics, music and memory, science and the arts, as well as telepathic communication, immediately elevate autistics into the giddy realms of creative genius that resides on the higher light frequencies of an open third eye.

Clairvoyance, clairaudience, claircognisance, clairsentience; Not woo-woo! Just terms used to identify gifts that were once considered phenomenal and downright scary to the uninitiated, but which have become more accepted over time due to their increasing prevalence. This is a prime example of how evolving belief patterns and their associated frequencies open up higher channels of communication at both an individual and collective level. But if we place these abilities alongside everyday occurrences within the animal kingdom, it can be seen that they're neither extraordinary nor phenomenal, but that the pineal is simply beginning to perform optimally and live up to its original design.

> *We've also witnessed groups of animals working as one mind. For example, when geese fly together, they do so in unison, communicating telepathically where to move... This is also a function of the pineal, which allows them to be in instant telepathy with one another. There's no individuality of thought, rather all are working as a collective whole for the welfare of the group.*

Andye Murphy, The Pineal Gland and the Third Eye Chakra, (2020)

Telepathy literally means *feeling from afar*, telling us that all the geese are feeling each other's energy, to ensure that they're operating on the same wavelength, so they move as one in the direction of flight toward their desired destination. An energetic navigation system with built-in trust. **[telepathy: see glossary]**

Now imagine if some of these geese lost their telepathic function. They'd be honking directions at each other... "turn left at the next tree", "slow down", "no, follow me, I'm the leader". There'd be utter chaos, a flurry of mid-air bingles, and very noisy skies. Those geese that were still tuned-in to their telepathy would feel the crazy lower vibes of their chaotic friends crashing in around them, disturbing their calm and distorting communications across the flight pack.

And now imagine that those tuned-in geese are the autistics, flying high on their telepathic wave lengths, navigating their way purely through light frequencies. They're filled with the joy and ease of unity, courtesy of their happy inner guidance system.

They feel connected to everything and everyone through the Earth's magnetic field. There's instant communication; easy, precise and crystal clear, as long as they're with others who are on the same wavelength. However, in this mainstream world, this is not the case, for they are surrounded instead by their honking chaotic compatriots, the fear-driven and emotional neurotypicals, who are lacking the telepathic function to connect to the collective whole; flapping their wings and issuing rules and directives that actually distort the clarity of flow, and make navigating in the world very difficult for autistics. Neurotypicals, like the chaotic geese, are still on L-plates in the school of telepathic navigation, and they're still searching, outside themselves, for their desired destination. Trust is not an in-built function. **[trust: see glossary]**

Evolution provides the opportunity for the chaotic neurotypical geese to hone their communication skills and enhance their connectivity through the gradual refinement of their telepathic function within the pineal...the opening of their third eye...simply through the expanding of the imagination. In fact, autistics possess their telepathic gifts only due to the evolutionary progression of their predecessors, who bravely pushed the boundaries and refined their own skills. *Everyone* is evolving and refining their telepathic abilities. *Feeling from afar* is a product of evolution. In fact, science is now attesting to the fact that we humans can sense more than we see through a magnetic 6th sense called magnetoreception.

> *It's part of our evolutionary history. Magnetoreception may be the primal sense...*
>
> *For much of the 20th century, magnetoreception research seemed as unsavory as the study of dowsing or telepathy. Yet it is now an accepted*

fact that many animals sense the always-on, barely-there magnetic field of Earth. Birds, fish, and other migratory animals dominate the list; it makes sense for them to have a built-in compass for their globetrotting journeys.

Mammals, too, seem to respond to Earth's field: In experiments, wood mice and mole rats use magnetic field lines in siting their nests; cattle and deer orient their bodies along them when grazing…

…tests showing that disrupting or changing magnetic fields can alter animals' habits. Scientists know that animals can sense the fields, but they do not know how at the cellular and neural level.

Science.org (2016), Magnetic Sixth Sense. Maverick Scientist thinks he has discovered Magnetic Sixth Sense.

The quiet, gentle, telepathic geese emulate the quiet, gentle, telepathic autistics, who are tuned-in to their own feelings and to those of others through magnetoreception. No need for honking or talking. In another surprising twist, being non-verbal aligns perfectly with telepathic communication, for it's simply not required. After all, sound does operate on a *lower* frequency, and therefore evolution is slowly but surely rendering its dense communication system obsolete. Can you imagine a world without all the noisy clatter? Well, it already exists on a higher frequency. Is it any wonder autistics are acutely sensitive to noise, when they operate from the telepathic stratosphere. **[noise: see glossary]**

> *...Henrik Mouritsen, showed that electromagnetic noise prevents European robins from orienting magnetically.*
>
> *Science.org (2016), Magnetic Sixth Sense. Maverick Scientist thinks he has discovered Magnetic Sixth Sense.*

When witnessed in animals and nature, telepathic communication is considered a wonder. But when seen in relation to human beings, it's considered woo-woo. Why do we put ourselves down? Why is it so hard to believe that humans can operate in the same way as our furry and feathery friends? It's just energy after all. Why is it not possible for *us* to be *wonder*ful? Full of wonder. Just like the animals. And just like infants and little children, who still see the magic in the world around them as they too ride the magnetic high vibes. And, believe me, infants *are* telepathic. Ask any mother.

I believe it is entirely possible for us to be *wonder*ful, and that autistics are leading the way. Autistics withdraw themselves from the dense external world of electromagnetism, so they can preserve their connection to the inner wonder and magic that's inherent in all of us from infancy; our original magnetic design. And in so doing they preserve the telepathic function of their pineal gland and they stop the fall into idiocy within themselves, thus contributing to the higher collective consciousness called evolution.

The pineal is returning to its optimal functioning, and autistics, as the forerunners of evolution, provide the physical evidence. However, this eccentric vision begs a new set of scientific spectacles in order to flip the currently held perspective that the autistic pineal is malfunctioning.

> *One of the proposed biological causes of autism is malfunction of the pineal gland and deficiency of its principal hormone, melatonin. The main function of melatonin is to link and synchronize the body's homeostasis processes to the circadian and seasonal rhythms, and to regulate the sleep-wake cycle.*

> T Shomrat, N Nesher (2019), Updated View on the Relation of the Pineal Gland to Austism Spectrum Disorders.

It would appear that the pineal has a direct effect on the body's homeostasis. At first glance, owing to low concentrations of melatonin and a prevalence of sleep disorders within autistics, it makes sense to conclude that their pineal is malfunctioning. However, when you don those eccentric spectacles, an alternative vision emerges, and it becomes obvious that the autistic pineal is operating like that of an infant. The autistic pineal gland is actually functioning more optimally, thus providing autistics with the same heightened telepathic communication that exists in infants...and geese.

What does this mean for homeostasis.

> *In biology, homeostasis is the state of steady internal physical and chemical conditions maintained by living systems. This is the condition of optimal functioning for the organism and includes many variables such as body temperature and fluid balance, being kept within certain pre-set limits.*

Homeostasis is brought about by a natural resistance to change when already in the optimal conditions.

Homeostasis, Wikipedia

Homeostasis, by virtue of its name, is a natural resistance to change, but this also implies a natural resistance to evolution. So, if what's accepted as natural, is that which has settled-in as familiar habit, homeostasis becomes the process of stubborn obduracy, where the body is obstinately holding on to the cycles of programmed memory...the pre-set limits. And its functioning is only optimal for the current set of environmental conditions. As evolution alters the environmental stimuli, it must also change the impact on the body and its resultant responses. Homeostasis must also evolve from one platform to the next-and-higher platform as the human body adjusts to the evolution of the environment. It stands to reason, therefore, that the evolution of homeostasis transpires as a direct result of the evolution of the pineal gland itself. Subsequently, differences in the pineal gland of autistics occur due to their being more highly evolved! They're standing on a higher platform, with infants and geese, sporting a more fully functioning pineal. Yet they're expected to respond like neuro-typicals.

Many spiritual practices believe that ancient humans had a fully functioning, open third eye, but that over time it closed as the pineal atrophied. This belief fits into a linear model of time providing the obvious perspective; that the closing of the third eye is a collective regression that has occurred over eons, and is therefore a function of *duration*. But it takes on a completely different perspective within the new framework of TIME, where the closing of the third eye becomes a function of *endurance* that plays out within the brain's limbic system throughout each individual's lifetime. With this understanding, the third eye closes under sufferance from prevailing environmental conditions, as discussed above, which directly affects the operation of the pineal. This would account for the variation in telepathic abilities within individuals, and make sense of all those spiritual leaders, through the ages, who have exhibited what appeared to be extraordinary gifts of insight. When evolution is seen as an individual pursuit, it allows for flexibility within the continuum of endurance, and provides a logical reason for the apparent genius of autistics. **[endurance & duration: see glossary]**

On hearing the word *ancient*, we usually conjure up an image of something that's very old and withered; however, its true meaning is actually *from before*. So, the understanding of *human beings from ancient times* can literally imply any earlier form of your own being that preceded where you are now. With the understanding that atrophy is caused through lack of nourishment, as its meaning attests, you can see that the pineal has simply been underfed and wasted away through lack of use in your earlier years. It's lost its true purpose over time, with *over time* referring to the progressive stages of aging endured by an individual throughout their life-time. With little nurturing, the fully functioning pineal of the infant, slowly atrophies on its journey through childhood into adulthood, thus creating a non-functioning monument to the defeat of your inner sight. **[ancient & atrophy: see glossary]**

The literal meaning of the word ATROPHY is *without nourishment*, or *wasting away*. But within this word, we can also see A TROPHY, meaning a prize of war, a sign of victory, a monument of an enemy's defeat. (Online Dictionary of Etymology) The pineal becomes a trophy of war instead of a portal of love. And the limbic system is the battle ground upon which the enemy of your inner sight becomes imprisoned by memory and emotion.

(In the following paragraph, I've italicised all the etymology meanings of the word *pine* from the Online Dictionary of Etymology.)

The PINE in PINEAL means yearnings, dreams and desires. The other etymology definition of the word PINE, being to *cause to starve*, *to languish* or *waste away*, aligns perfectly with the undernourishing of the pineal gland through childhood, which leads to it being *consumed with grief or longing*. This is felt, from the child's perspective, as *punishment* or the *penalty* for growing up, which causes *pain, torment, affliction, suffering,* an *enduring penance* called adulthood. Their dreams waste away in the depths of their being, taking on the new mantle of desires as they *yearn* to once again express their true essence. **[pineal 4: see glossary]**

The pineal atrophies during the growing up process that takes place from childhood through to adulthood, when the individual shifts their allegiance from internal to external guidance; from the imagination to cognition; from an open third eye to dualistic human sight; from the higher light frequencies of the creative field to the lower light frequencies of environmental light, called the visible light spectrum, and sound. The pineal atrophies as its light grows dim.

> *Environmental light acts through the retina and entrains the pineal gland's circadian rhythms by way of the hypothalamus and sympathetic nervous system. Light depresses the pinealocyte activity.*
>
> Frances Booth (1987), *The Human Pineal Gland: A Review of the Third Eye and the Effect of Light.*

Environmental light, being the electromagnetic field of emotion, depresses the activity in the cells of the pineal. The visible light spectrum forces the pineals true function of telepathic communication, dreaming, visions and the imagination into submission through emotional bullying. Emotions atrophy the pineal, as its spirits are depressed. Emotions are the trophies of your childhood battles as your infant essence goes into hiding.

You may remember, from Chapter Three, that the light photons of the electromagnetic field are affected by your emotions, which literally creates the world around you, your reality. So, the environmental light that informs your pineal, via the retina, is emotion-based. The *retina* (meaning *net*) captures the emotional cargo of the electromagnetic field in its net, from where it's transferred to the pineal and then transduced into a chemical memory (emotion) within the endocrine system. The optimal functioning of the pineal, which can only take place on the higher light frequencies, is depressed under the weight of heavy emotions. And yes, there's a definite link here to depression. The electromagnetic field of the visible light spectrum, wherein humanity plays, is a sad blanket of emotion that contributes to the atrophy of the pineal gland, and the closing of the third eye. It's this sad blanket that's being lifted from the innate functioning of autistics thanks to their predecessor's courageous foresight. However, the lack of higher understanding that's woven into the sad neuro-typical blanket of emotions can create enormous frustration for autistics, leading them too into depression because they're simply not seen for who they really are.

Thankfully, evolution is indeed fostering the progressive opening of the third eye through the gradual lightening of emotions as individuals make conscious feel-good choices from their inner guidance system, thus imbuing the cycle of light with increased brilliance and creating a new world reality that's increasingly inspired by love. Through a greater emphasis on the nurturing of *self*, the pineal is bathing in higher light frequencies and opening the portal to the magnetic field of creation that offers unlimited possibilities; the field of creative genius; the first and fondest memory of the joyful infant. For, nurturing is creative...not destructive. Nurturing feeds the pineal loving, creative vibes. The pineal is the portal to the imagination. And autistics are privy to the magic that lay within. Autistics are forcing humanity to lift their game and be living examples of love and kindness, and in order to do that they need to re-ignite their imaginations. **[nurture: see glossary]**

Because the pineal is such a delicate topic, owing to its long-held spiritual associations with the third eye, I'm going to employ the exquisite fairy tale, Pinocchio, to further illuminate this explanation. Fairy tales have a way of light-heartedly illustrating complex issues with crystal clarity and wonder, which is why they're a favourite among autistics who intuitively connect with their deep, hidden meanings. And selfishly, it's because there's such a delightful pineal/Pinocchio connection that I'd love to share it with you.

Let's start by looking at the name PINOCCHIO. It's formed by joining the word PINE (which means to yearn, dream or desire, and is a type of wood) with the Italian word OCCHIO, meaning eye. The PINE EYE, by its association with the third eye, is none other than the pineal. The little wooden puppet, Pinocchio, is representative of the pineal gland (the portal to the imagination). And if we follow the fairy tale from its Once Upon a Time to its Happily Ever After, the pineal reveals itself in all its evolutionary glory, as Pinocchio transforms from a little wooden-head into a real boy.

But first to an interesting revelation that's crucial to the plot. Although the obvious function of the pineal relates to the transmission of light frequencies via the retina in your eyes, it also has a less conspicuous relationship with smell. The brain's sensory system that processes smell is called the olfactory system. It communicates directly with the limbic system, which is wired to the pineal gland, creating a discreet, but powerful, connection with smell. Both the pineal and Pinocchio share this smell connection. I'll let Pinocchio explain. **[olfactory 1 & 2: see glossary]**

We all know that when Pinocchio tells a lie, his nose grows. But with a little twist, the word nose can be substituted for bouquet, aroma or smell. A posy of flowers is called a *nose*gay; a perfumier is known as *the nose*, and wine tasting employs the term *nose* as the smell or bouquet of the wine. So, the growing of Pinocchio's nose is a beautiful play on words that represents an increase in smell. When Pinocchio tells a lie, he becomes smelly...*something doesn't smell right!* He's not following his truth, which subsequently pulls him further from his dream of becoming a real boy.

By applying the pineal analogy, we discover its function and subsequent effect on the physical body in relation to smell, and further to the light frequencies of your reality. For when you don't express your truth, your pineal deposits a concoction of smelly chemistry within your cells; the emotion that drives your behaviour and creates a reality that doesn't match your dreams. By association, because the pineal is not being informed by your highest truth, but instead by the dictates of the external world, your third eye closes to your heart's desires and your pineal falls into a state of yearning. It's sad. It's pining.

The pineal is the eye through which you animate your dreams and desires. Temple Grandin's thinking in pictures is a fine example, as she imagines all her design solutions with her inner sight, thus breathing them into her reality. Her design is a fully formed image; all she need do is believe in it enough to transfer it into the external world; to be inspired enough by her imagined design that its essence animates her cells, thus motivating her into the action of taking steps toward its creation. Whatever you can imagine, you can create. From inner to outer. The pineal, therefore, is the portal through which you connect to the inner field of creation, the wondrous imagination, where all possibilities already exist, awaiting animation.

The Online Dictionary of Etymology informs us that to animate a body is to endow it with a particular spirit. It is spirit, therefore, that holds the magic wand of creation. Spirit is *your* essence, *your* obliging genie or granter of wishes, *your* creative genius, *your* self-belief that dips into the unlimited depths of imagination and breathes to life that which inspires *you*. Inspiration is the work of your *inner* spirit...and that spirit manifests as smell. Your imagination has a magnetic essence, which expresses through your pineal in the form of smell. The aromatic essence of your creative spirit literally animates your dreams into reality through your pineal. You can animate anything into existence simply by believing in, and nurturing, it. **[magnet & inspiration: see glossary]**

Inspiration exists but it has to find you working.

Pablo Picasso

Your cells are little oil pots containing the essence of your imagination...your spirit of belief...your magical genie...your granter of wishes and dreams...your smell. Tucked within the heart of every cell is your unique aromatic signature in the form of your chromosomes, otherwise known as your DNA. It's this unique aromatic spirit that animates your cells to create your body and your reality. And it's this unique aromatic spirit that's refining its essence throughout evolution, and thus animating a higher reality for humanity. When you cross the evolutionary finish line, your cells will be fully refined and cleansed of all their smelly prisoners, chemistry, emotions...DNA. **[gene 1, 2 & 3, chromosome, chemistry & function: see glossary]**

> *Smell is the only fully developed sense a fetus has in the womb, and it's the one that is most developed in a child through the age of around 10 when sight takes over.*
>
> *And because "smell and emotion are stored as one memory", said Goldworm, childhood tends to be the period in which you create "the basis for smells you will like and hate for the rest of your life."'*
>
> Colleen Walsh (2020), What The Nose Knows

It seems obvious to me that smell would be fully developed in the fetus, considering it's the very nature of chromosomes.

Evolution, therefore, has a direct relationship with smell. The human race appears to be a relay through time, where the aromatic baton of chromosomes is passed from one generation to the next, continually refining its essence on a pilgrimage toward purification. We're following the trail of scent that's been left by our ancestors. As forerunners in this evolutionary relay, autistics have been handed a highly refined aromatic baton, leaving them susceptible to the smelly lingerings of the neurotypical majority, which is why they're so sensitive to smell and prone to withdrawing from the density of the world.

> *Autism is one of the most heritable genetic conditions we know about.*
>
> Professor Andrew Whitehouse (2018), *National Guideline for Assessment & Diagnosis of ASD in Australia*

This also confirms, yet again, that evolution is an individual pursuit, with every human being situated along a continuum of transformation, in accordance with their creative essence. Evolution and the olfactory system are intimately connected, and both express through the pineal, courtesy of your limbic system. **[evolution 3: see glossary]**

Our fairy tale equivalent shows us likewise, that both transformation and smell express through Pinocchio on his journey to fulfilling Geppetto's dream that he become a real boy. As the creator of the dream, Geppetto represents anyone who's ever wished for a dream to come true. For we are the creators of our dreams, and evolution is honing our skills of creation and leading us toward that very outcome…all our dreams come true.

When we first meet Pinocchio, he's a little wooden-headed puppet. A beautiful illustration of the pineal in its non-creative state of atrophy, functioning instead from the rigid habits of childhood programming, which subsequently control the strings and joints of the body. We become puppets, pulled by the conditions of the external world having traded our imaginations for rules, our power for approval, our flexibility and youth for stiffness and aging. But then the Turquoise Fairy, who's indicative of your intuition, offers a spark of light, by reminding Pinocchio that it's up to him to make Geppetto's dream come true. He has the power, but he must learn to listen to his inner voice instead of the temptations of the external world. Your pineal has the power to create your dreams when it's in tune with the illuminations of your sparkling creative spirit rather than the dull rays of the visible light spectrum, heavy with emotion.

The gift of life is bestowed upon the little wooden-headed Pinocchio, and Geppetto is overjoyed. Likewise, your pineal has the gift of animation, illuminating your body when you present it with the highest essence of your creative spirit. In other words, when you're happy, your body tissue regains its youth. As an infant, this was your joyful blessing, the first and fondest memory gifted to your body tissue. This is the creative realm that autistics wish to hold on to; their inner sanctuary; their vivid imagination. Unfortunately, the imprints of your life experiences within the external world soon begin to bury this joyful gift beneath layers of protective emotion, disconnecting you from your true creative essence. It's still there, calling to you from the depths of your soul, awaiting re-animation. But in the interim, your pineal processes the less-than-creative programmed memory that's not your truth...your pineal is telling lies...and its *nose* is growing, as it sends the smelly soldiers into your cells in the interest of self-preservation; all triggered by fear. **[intuition: see glossary]**

And so it is that Geppetto introduces the external influences to Pinocchio by sending him to school so he can learn the ways of the world. However, the tempting, worldly fox, representing the temptations of the external world, cunningly leads him astray with enticing stories and promises of future reward. Mesmerised, Pinocchio innocently follows the tantalizing trickery. The pineal is blissfully unaware of its new role in the theatre of life, acting to the script of others in the process of growing up, while cunningly settling the mask of your persona over your true essence. For Pinocchio, it soon becomes clear that the invisible strings of external control are establishing themselves within his new life. He's captured in a cage of his own creation. For the first time, Pinocchio expresses emotion. The pineal expresses the chemical reactivity that beds down as emotion in your cells when you're no longer free to express your truth. Your spirit is sad. In the midst of all this emotion, Pinocchio wants Geppetto, but Geppetto has lost sight of his dream, and is searching hopelessly in the dark external surrounds.

Your pineal wants to reunite with your creative spirit, and your creative spirit wants to find your pineal. As creator, all you want is to embrace your first and fondest memory and open your pineal to the lost joy of your pure inner light, but you're in the dark and can't find your way home. **[infancy: see glossary]**

The talking Cricket steps in to save the day...well, perhaps not, but his intentions are honourable. The chatty Cricket represents Pinocchio's conscience; the dualistic, finger-wagging, rule-keeping memories of the past that constantly vacillate between right and wrong. But no matter how hard he tries, he simply doesn't have the solution to release Pinocchio. The instinctive ancestral voices of approval and disapproval that chirrup incessantly in your mind, will never set your pineal free. Only your gentle inner voice of intuition, the Turquoise Fairy, is able to guide you back to your truth. Your true guide is always gentle, nurturing and kind. She magically appears, and Pinocchio is delighted to see her, but her presence reminds him that he's strayed from his promise to be brave and true and unselfish. He thinks he can get away with it by embellishing his situation, but as the truth stretches, so too does his nose. You can't fool your intuition. Pinocchio is horrified, but intuition shines through once again as the kindly Turquoise Fairy forgives him, and his nose regains its integrity. Your body nurtures its pure essence when guided by your intuition. Forgiveness is the blessing that cleanses the pineal and resurrects the physical body, reinstating purity and integrity within its fibres. **[Talking Cricket & Turquoise Fairy: see glossary]**

As this common English idiom states,

It's as plain as the nose on your face.

The more you disconnect from your true essence, the more your emotions grow and grow, until your cells are so smelly that disease physically manifests in your body tissue, so you can no longer ignore it. The issue is in the tissue...literally! But each time you intuitively guide yourself back to your true creative essence, through forgiveness, you refine the smelly chemistry of your pineal and continue the pilgrimage back into the pure essence of your first and fondest memory. Truth, expressed through the pineal, brings integrity to your body tissue.

Unfortunately for Pinocchio, lessons are hard learned, as temptation proves irresistible yet again, luring him to the Land of Toys, where he can be one of the boys, enjoying the pleasures of adolescence. However, all these pleasurable treats come with a side-serving of pain. We suddenly find that the Land of Toys is in fact the pleasure/pain centre of the brain...the limbic system, with all its programmed memory of ancestral dictates. By foolishly following the desires of others, the boys are turned into donkeys, becoming beasts of burden carrying the heavy emotional load of adult obligation in the form of hormones. Puberty turns the pineal smelly." **[adolescence: see glossary]**

The cycles of life see each new generation of youth slowly waking-up to the hidden dangers of this promised land, where their childhood magic disappears into smelly bundles of emotion-laden hormones. Only by diving deep into the emotional baggage can they clear the misunderstandings of past beliefs, allowing them to own their unique inner voice and express their truth with beauty and wonder. And so it is that Pinocchio makes haste to escape the clutches of his own donkey-nature within the deep troubled waters of emotion, emerging with the awareness that it's time to go home. It's time to go back to Geppetto, his creative truth. Your pineal recognises that the temptations of the dualistic external world, land you squarely in the dualistic island of the brain's pleasure/pain centre. Evolution has awoken the pineal from its blind sleepwalk of unconscious smelly programming, into a new dawning of realisation that its original essence is the only true expression.

Autistics are acutely aware of the dangers of the tempting promised land, due to their more highly refined aromatic bundle of chromosomes, gifting them with an acute sense of smell. They sense the pending arrival of their donkey-nature, as it descends upon their innocent spirit. And they escape its clutches by withdrawing from the world, so they can stay tuned to their unique creative essence, in an attempt to save their magic. This is why their limbic systems are less active. But it isn't easy in this sleepy neuro-typical world that hasn't yet fully evolved.

Pinocchio discovers that Geppetto, his creative spirit and truth, has been swallowed by a monstrous sleeping fish, so he endeavours to liberate him from the deep waters of emotion that are fraught with fear. Your creative spirit and truth have been swallowed up by the neuro-endocrine system of fear and emotion, where they sit sadly inside the monstrous lie that tells you you're not good enough. Within this monstrous illusory fabrication Geppetto is starving, a testament to the under-nourishment of your hidden creative spirit, your imagination that's been suppressed, the loss of your true essence of beauty and wonder. For with the death of your creative spirit, your truth shrivels, your life depletes, and your animation withers and dies. Mortality sets in.

The monstrous lie that the sleepy neuro-typical world perpetuates is that we are mortal beings of so little worth that we have no power over our own bodies. For science believes only what it sees, because it's forgotten the spirit of creation with its magical ability to animate whatever we can imagine and believe in. Evolution, however, defies science again and again as it slowly but surely rouses the slumbering third eye. Shifting the monstrous lie of mortality relies upon the delicate process of refining the aromatic signature of each successive individual throughout the process of evolution. Smell awakens the sleeping monster. Perfume is evolution's catalyst.

Your creative spirit (smell), and your pineal have been working together throughout evolution to refine the aromatic signature within the cells of humanity, animating your body tissue with increased purity and youth. Geppetto and Pinocchio epitomise the process of evolution that is creating increased longevity within the fibres of the human body. And when the process is complete, Pinocchio will be a *real boy*, for your pineal will express the *integrity of youth* through your physical body. **[youth: see glossary]**

Geppetto and Pinocchio are essentially clearing the smelly prisoners, the sad memories, the fearful defensive memos, the olfactory notes that are stored in your limbic system as the limiting stories you believe about yourself and your life, which stop you from manifesting your dreams. Geppetto and Pinocchio are assisting humanity in opening to a new reality by escaping the limbic system with its monstrous burden of emotional family history. Geppetto and Pinocchio have put up quite a struggle.

But transformation and rebirth are always preceded by a struggle with, and death of, the old, making way for renewed life. Evolution requires the death of your rigid, under-functioning pineal, so it can reinstate its true function as the open doorway for your guiding light, your intuition, your creative imagination. And so it is that the Turquoise Fairy's magic light transforms Pinocchio, turning him into a loving youth, because he showed integrity, courage and selflessness in saving Geppetto. It's your intuition that restores your pineal to its optimal functioning, as you rescue your creative essence from its sad prison of limitation, and invite it to frolic freely in your open field of imagination, where all your dreams await. When you follow your heart's desires, trust what you feel and express your truth, you shine your light through the fabric of your being.

Your intuition is the bright and beautiful star that makes your dreams come true. So, when you wish upon a star, you're holding your dreams in the highest light, nurturing them in your heart, and believing them into reality, through the portal of your pineal.

When looking at the pineal, specifically in relation to autism, it's role in evolution becomes obvious.

Your pineal is the projector through which your life movie manifests as your reality. It casts your imaginings onto the blue screen of planet Earth. Evolution has seen the pineal opening to increasingly brilliant imaginings because the creative inspiration and intuitive whisperings of successive generations boldly questioned and dissolved the boundaries of limitation. Physically, this means that the limbic borders of the brain's third ventricle have been clearing their limiting memories. The third ventricle is opening to higher possibilities, higher imaginings, a higher flow of creative essence through the pineal...the third eye is opening. The process of evolution *is* the third eye opening. And it leads to a condition that's currently called autism. Clearly, autism is not a disability, but a higher ability that's simply misunderstood, because this simple truth sits in the neuro-typical blind-spot.

If we align Pinocchio's journey with that of evolution, we can very neatly place autistics toward the final stages, where their pineal glands are more optimally functioning, as they're guided by their hearts into a higher vision that reflects higher integrity; truth. Evolution is transforming humanity from being little wooden-headed puppets...pulled by the strings of the external world...into being real, and thus guided by true inner feelings. I propose that the autistic's pineal gland provides evidence of the so-called spiritual gift of inner sight, which is far from the malfunctioning organ of its scientific diagnosis.

Through the portal of the pineal, evolution is bringing spirit and science together in an harmonious dance of energy and matter; a nurturing of nature. For as you listen to your intuitive whisperings and nurture your spiritual imaginings, they express through your physical being to manifest as your reality.

Autistics walk easily through this portal of the imagination into the playground of unlimited possibilities, where magic turns dreams into reality; where wonder and beauty abound.

Evolution is a joint spiritual/physical transformation from dis-ease to ease: from reactivity based on past information that's stored in your memory banks, to calm in-the-moment clarity of choice; from emotion to feeling; from the limited programming of a rigid brain to the creative expanses of the imagination; from an external to an internal focus. Vision is shifting from the lower light frequencies of the visible light spectrum that's inherent in dualistic human sight, to the higher light frequencies of the open third eye, the mind's eye, inner sight, making your sweetest dreams come true. **[evolution 2: see glossary]**

Love, nurture and massage your secret longings. Live them, breathe them and feel them, for only when you match the wonder and beauty of your sweet dreams, will they come to life. **[fate: see glossary]**

This is my favourite Albert Einstein quote...

> *Everything is energy and that's all there is to it. Match the frequency of the reality you want, and you cannot help but get that reality.*

It can be no other way.
This is not philosophy. This is physics.
Albert Einstein

Chapter Seven: The Light at The End of The Tunnel

...blossoms smell like bodies

The proverbial light at the end of the tunnel indicates that a long and difficult piece of work is finally reaching its conclusion. In perfect analogy, we find that the pineal lights up, a beacon of truth, following the progressive refinement of smell within the cells of humanity, in the lengthy and difficult piece of work called evolution. **[flower: see glossary]**

> *The idea that flowers smell like bodies, of course, seems like a strange proposition. But scent is an amazing thing,*
>
> *and the science is unequivocal...*
>
> *the connection between scent...and skin is always there as soon as we look beyond the surface.*

Tilar J Mazzeo (2010), The Secret of Chanel No. 5 **[skin: see glossary]**

Lingering beneath the surface of your own skin, sits the unique bouquet of your personal chemistry; the aromatic electro-magnetic arrangement of your favoured emotions, which determines the very function of your cells and the subsequent form of your body matter. This is made very clear in the word *matter* itself. I usually prefer to assign any word break-downs to the glossary, but feel that its presence here will serve to clarify my meaning in this instance, and illustrate just how intimate a relationship your body tissue has with smell.

MATTER = M + ATTAR

Attar: perfume of flowers (especially roses)

f. Persian 'itr, f. Arabic 'itr = perfume, aroma

MATTER = (M)/EM + ATTAR = ElectroMagnetic **ATTAR**

MATTER = ElectroMagnetic floral perfume/aroma esp. roses

Bodies smell like blossoms. Mother nature deems it so.
Your chemistry, your DNA, your chromosomes *are* smell. Your internal code of communication is conveyed by smell. Your body matter is glued together by smell. You are a temple of smell; a living, breathing cellular aromatic garden. **[chromosome + romance: see glossary]**
The following quotes reiterate this fact, although you'll see that scientists haven't as yet completely joined the dots.

There's a large genetic component to body odor.

and this...

Scientists still don't know how human body odor can act like a scented fingerprint.

and this...

A hefty pile of evidence suggests that emotions have a scent.

So, it's your scented emotions, sitting in your cells and organs, that determine your body odour, the genetic, scented-fingerprint of your body matter; casting your very essence into the electromagnetic field to create your reality. The electromagnetic field, being a field of emotion, is therefore a big bouquet of scent.

And this final quote must certainly help in the understanding of autism, and support my theories around its association with smell and the limbic system; as well as emotions.

When we catch a whiff, the areas of the brain responsible for social processing light up

The above four quotes are from:
Marta Zaraska (2017), The Smell of Scent in Humans is More Powerful than we Think. Discovery Magazine

With the ultimate roots of the word *social* meaning *follow*, it's clear that we literally follow our noses in social situations, and that our aromatic fingerprint is our guide. If autistics *are* more highly evolved, as I propose, then so too is their aromatic fingerprint. Their apparent deficits in social processing would indicate a more highly refined bank of olfactory triggers and guides, wafting from a more beautifully scented internal garden. Concurrently, this would render them acutely sensitive to the unpleasant end of the odour spectrum, causing them to withdraw from the smelly chemistry of the neurotypical world in order to protect their highly refined olfactory mechanism and sensitivities.

We're told that...

Autism is a developmental disorder characterised by difficulties with social interaction and communication.

Autism Spectrum, Wikipedia

However, with the understanding that the word *social* stems from the Latin *sequor/sequi* meaning *to follow*, our definition now tells us that autism is characterised by difficulties in *following* the expectations of the mainstream world.

Autistics lack the steel to follow the dictates of those with a less refined aromatic fingerprint.

But autistics are forerunners in the evolutionary progression into higher intelligence, which places them *ahead* of mainstream understanding.

Autistics are leaders...not followers.

Their more highly evolved aromatic fingerprint affords them a more highly refined processing mechanism, which is powered by the internal motivation of their heart. Autistics are highly inspired vessels of innovation.

Autistics are shining their bright light upon humanity and inviting neurotypicals to join them in their delightful gardens of wonder. For that's the goal of evolution for each and every one of us; to go within and resurrect our individual essence, and gather as one big beautiful bouquet of ingenuity. **[garden: see glossary]**

So, yes, bodies smell like blossoms, and when we reach the light at the end of the tunnel, having concluded the evolutionary refinement, bodies will smell like roses, the most highly purified floral essence, epitomising love and beauty. A romantic notion, I agree, but indisputable, as it's written into the word itself. **[matter: see glossary]**

The proverbial light at the end of the tunnel could just as easily be the happily ever after of your favourite fairy tale, where everything's coming up roses. Essentially, having traversed an extended period that's been fraught with difficulty and unrest, a glorious resolution is in sight, bringing relief and great joy. Throughout the journey, innovation paves the way for transformation, leading to a conclusion that bathes you in the light of a new and higher vision; a new and higher reality.

Evolution is leading humanity to the brilliant light at the end of its most enduring tunnel, through cycles of progressively higher vision called generations. In the fairy tale of evolution, the increasing cycles of higher and higher intelligence *within* each of those successive generations, has contributed to a world with brighter and brighter prospects. And dotted amongst them are the sparks of pure genius that effervesce within the minds of outstanding individuals...the forerunners. With each generation the light gets brighter, as the higher vision from the previous generation is bestowed upon them, awaiting even further illumination. The overall intelligence and vision of humanity is, therefore, steadily climbing. Inevitably, the number of genius individuals dotting our planet is also rising incrementally. However, without the understanding of evolution's plan, their shining lights can appear utterly blinding, creating bewilderment and fear within the mainstream world...and the resulting misnomer, autism.

Autistics are actually leading us toward the light at the end of the tunnel; to evolution's brilliant resolution; to the rosy garden of delight that awaits us all...within our own bodies. In this chapter, I'm going to dive, once again, into the inner workings of the physical body and reveal the microcosmic evolution that parallels the macrocosmic cycles of increased intelligence and higher vision. I'll also pinpoint exactly where *the light* and *the tunnel* are situated within each and every individual, awaiting reconnection. And I'll reveal their relationship with the pineal, the limbic system, smell and autism.

I'll also be calling on another favourite fairy tale to bring clarity and a light heart to what could otherwise become a stodgy explanation. Perhaps someone as white as snow, who has seven diverse little friends. But first, let's take a look at the word *light*, for it has an interesting array of applications. And with a slight tilt of your perception, you'll be able to see it's connection to smell; and its intimate relationship with autism.

Light:

> *Brightness, radiant energy, that which makes things visible; illuminate, ignite; having little weight; easy to do; shining, bright, beautiful; white, moon.*
>
> *Online Dictionary of Etymology*

The most obvious example of *light* or *that which makes things visible*, is the visible light spectrum (VLS), which is the spectrum of electromagnetic radiation that is visible to the human eye. Interestingly, the frequency parameters of the VLS are not rigidly fixed, as they may vary within individuals; this is our first clue that vision is mutable and dependent upon individual perception. The second clue rides on the knowledge that many species of animal can perceive beyond the VLS boundaries.

So, what comes to light for *you* as an individual (i.e. your reality) depends upon *your* specific vision; your unique perception; the story you hold true; your store of beliefs. This is your personal nest of light frequencies that you tune into with ease and habit like a favourite radio station. Therefore, *that which makes things visible* is your belief system. Vision equates to beliefs therefore light equates to beliefs. You cannot bring to light that which is not massaged within your cellular bundle of beliefs. In other words, what you *believe* becomes what you'll *see*. Your beliefs determine the measure of light that literally shapes your physical reality. And as you stretch your beliefs through the power of your imagination, you massage a new, brighter reality into being; one that is beyond human limitations.

Your imagination is the most powerful force in the universe. It's magnetic, and attracts your reality to you.

Imagination precedes, and prescribes, reality. Imagination is your creative motivation.

Let me reiterate my favourite Einstein quote, for extra emphasis:

> *Everything is energy; that's all there is to it.*
>
> *Match the frequency of what you desire, and you cannot help but manifest it.*
>
> *This is not philosophy. This is physics.*
>
> *Albert Einstein*

Any child will tell you that your imagination is real, and that it always comes true. But before tackling the imagination as creator, I'd like to revisit the connection between visible light and emotions, in relation to beliefs. As discussed in an earlier chapter, the light photons of the electromagnetic field are affected by your emotions, therefore visible light is infused with emotions. So, *that which makes things visible* is infused with emotions. Your reality is literally the way it is because of your stored emotions; the electromagnetic ocean of beliefs that you store in your cells in the form of smell.

The electromagnetic light field is a blanket of emotion that holds humanity's beliefs in the form of smell.

To reiterate one of my previous quotes:

> *A hefty pile of evidence suggests that emotions have a scent.*

It stands to reason, therefore, that light and smell have an intimate relationship. If you re-imagine your beliefs, thus refining your emotions, you lighten the smelly load that's carried by photons in the electromagnetic field. You literally purify the atmosphere. (The cleansing of the planet begins from within! Imagine how *this* affects global warming!) Your imagination, as creator, fosters increased light, both in the light field and in your cells. With each re-imagining, your body is primed to receive higher light frequencies. And so, the light cycle continues, with the newly-primed photoreceptors of your retina transferring its lighter aromatic cargo to your pineal, to be transduced into a lighter chemical memory (emotion); a more refined smell; in alignment with your new higher belief. Your higher vision creates a higher vision. And the cycle continues.

Your imagination, as creator, refines your chemical memories or smell. Therefore, your imagination refines your chromosomes. Imagination defines your aromatic fingerprint (DNA).

Imagination inspires evolution.

The blossoming of your powerfully magnetic imagination is leading you to the brilliant light, and rosy bouquet, at the end of the evolutionary tunnel. And as fore-runners, autistics are a bloomin' inspiration. **[imagination: see glossary]**

This leads directly into another of the definitions for light, being to *illuminate*. The Oxford Dictionary of Etymology explains that to illuminate means to *enlighten spiritually*. With the understanding that your spiritual essence is smell, it becomes clear that evolution, through the creative pursuits of the imagination, is progressively purifying your body essence or smell, thus illuminating your body tissue with an aromatic fingerprint that's infused with increased lightness. In so doing, evolution is illuminating your body tissue with increased youth.

Your body is the living, breathing architectural personification of your imagination; its blueprint being smell.

Now, doesn't *that* change the perception of the autistic body with its bespoke architecture.

Evolution is the movement through time that simulates the blossoming imagination, creating an increasingly fragrant garden within the fabric of humanity.

> *Our bodies are gardens to which our wills are gardeners.*
>
> William Shakespeare, Othello

and this,

> *...to scientists, there is a simple reason why humans are so powerfully attracted to the scents of certain flowers; whether we know it or not, blossoms smell like bodies.*

Tilar J Mazzeo (2020), The Secret of Chanel No. 5 (page 77)

Your open pineal embeds your blossoming imagination in the fibres of your being.

In the previous chapter, Pinocchio and his fairy tale family revealed the workings of the pineal gland in relation to smell and the limbic system. Let's now pick up from the pineal and follow the scent as it trails through your body contributing its magnetic essence to your architectural form and function. **[function, olfactory and evolution 3: see glossary]**

Every function of your body is determined by smell, which is planted within your cells. Your olfactory system of smell is like an oil factory, as it manufactures your body's essential oils and packages them in your little cellular oil pots within the fibres of your being. As evolution refines your essence, it also refines its function; how it manufactures your physical structure. Your unique essence is very powerful.

Therefore, the portal of the pineal has a powerful responsibility as she sends the scented chemical messages into the organs of your endocrine system, which align with the energy centres that are situated along your spinal channel. Your energy centres are called chakras. Once again, please don't baulk at the spiritual term *chakras*, for they too have a physiological counterpart, which I'm about to explain.

Can I also remind you of Einstein's words that *everything is energy*.

The word *chakra* literally means wheel or vortex, because it looks like a spinning disk of energy. Chakras are situated along your spine, aligning with your major endocrine glands and nerve networks. Already there's a connection between your energy centres and the neuro-endocrine system, and accordingly with the aromatic fingerprint of your internal garden; your DNA.

> *Each chakra holds the beliefs, emotions, and memories related to specific areas of our life.*

Parita Shah (2020), *A Primer of the Chakra System*. Cited by Chopra.

Your chakras cannot be seen with the human eye that operates in the dense, emotion-laden visible light spectrum, which is why it's so hard to believe in their existence. You can, however, sense or feel these energies, and see them with your mind's eye. This is your intuition coming into play again; your Turquoise Fairy communicating with you; gently nurturing and guiding you via the energetic system of sensations within your own body. Everyone has the capacity to tune into their energetic system with a little practice, by simply slowing down, becoming still, and *feeling* it. Autistics do this naturally. They feel everything.

Your body is always trying to look after you by telling you how she's feeling. She's trying to get your attention so you can recognise any *dis*-ease and make nurturing changes before it settles as *disease* within your body tissue. Your body is urging you to re-imagine your beliefs so she can feel at ease. And when you do, you begin romancing your body by handing yourself a beautiful bouquet of flowers. By so doing, *you* dictate the condition of your body tissue. **[disease & tissue: see glossary]**

'Tis *you* in every tissue.

'Tis *you* who fills your body tissues with your unique energy and essence. It's *your* story that designs, creates and evolves every tissue in your body; every fibre of your being. 'Tis you that radiates from *your* body, creating *your* world. 'Tis you who determines your body's form. 'Tis you who determines the functioning of your body. 'Tis you within your aromatic fingerprint! 'Tis you. I'm sure you get my point!

So, the energetic communication system of your body, reflects the state of your neuro-endocrine system, and provides a sensitive feedback loop that's perpetually guiding you; your very own nurturing Turquoise Fairy; your intuition. Intuition is simply your body giving you information, which guides and nurtures you and your behaviours. **[intuition & life: see glossary]**

Autistics listen to the energetic intuition of their Turquoise Fairy; their body's sensitive feeling system. They are exquisitely sensitive because they tune into the higher frequencies that are all around them, just like the telepathic geese in Chapter Six. Their intuitive whisperings are of wonder and beauty, kindness and love; of magical patterns and higher intelligence; of the gentle and ingenious qualities of their innate brilliance. Their essence is pure. So, when the energies of the people and experiences of the world don't match the autistic's higher energies, they feel it acutely. And they shy away. They're not *too* sensitive; they're *beautifully* sensitive. Just like you...when you tune in, and romance, your body.

Autistics are showing you how to listen to your body's energetic chakra system. Let's take a little look at what that means.

There are seven main chakras sitting along your spinal channel from tip to top. Each chakra is associated with a different colour and frequency as well as particular endocrine organs and nerve networks. Here's a general summary.

Base Chakra, red, the adrenal glands.
Sacral Chakra, orange, the reproductive organs.
Solar Plexus, yellow, the pancreas.
Heart Chakra, green, the heart, thymus and circulatory system.
Throat Chakra, blue, thyroid and hypothalamus.
Third Eye Chakra, indigo, pituitary and pineal.
Crown Chakra, violet, upper brain/crown.

You may have seen images of Kirlian photographs showing these colours of the rainbow sitting like traffic lights along the central channel of a human figure. The Kirlian photography is designed to collect the energetic data that's flowing through your body's network of subtle energy channels called meridians, which indicate how well your organs and nervous system are functioning. This information gathers at your chakra centres. When your emotional, mental and physical states change, so too does your energetic readout, which is indicated by subtle variations in the intensity of colour.

> *Of course, each chakra is related to a specific part of the body as it pertains to your physical health. By the same token, each one of the seven chakras is related to certain emotional or mental health characteristics.*
>
> *Depending on whether or not each chakra is open or blocked, you may experience corresponding emotional issues as well as physical symptoms.*

Jacob Oleson (2013), Colour Meanings: The 7 Chakras and their Meanings.

Your emotions, which are in fact re-enactments of old memories, settle into your body organs, depleting their energy and hindering their optimal functioning. Emotional *dis-ease* manifests as physical *disease*, promoting decomposition and increased smell. Your inner garden is wilting.

It's my belief that the colours radiating forth from the body, as seen in Kirlian chakra photographs, is actually emotional energy that's *leaking* from the body organs, indicating their varying states of dis-ease and disrepair. The subjects of these colourful light maps, whether old or young, are human beings who've been impacted by many life experiences, which have inflicted wear and tear upon their body parts in their repetitive retelling, affecting both the form and function of their mechanical apparatus. Old memories create aging, and a loss of power.

You literally *give your power away* to the external world when you comply with the demands of the past, instead of listening to how your body is feeling in each situation and responding in a way that nurtures it...and you. Whenever you respond emotionally, you've lost power; the world around you has disempowered you; your body organs are under-functioning and aging, and your energy centres are leaking. The surrounding electromagnetic field is an energy thief, drawing its power from your leaking emotions.

The Kirlian photographs are a wonderful tool for discovering the specific areas of your body that need nurturing, and the accompanying emotional triggers that need to be addressed. They guide you to the situations where you need some re-imagining of your current story or set of beliefs.

With their less emotional inclination and higher belief system, I imagine the natural aura of autistics tends toward being clearer and brighter. They choose to nurture their bodies by withdrawing from the density of the external world so they can tend to their internal garden, where they feel more at ease. Ease promotes power. And more power means brighter lights, increased clarity and a higher vision. So, the colours of a life turned inward, are clear and bright.

The separated colours of your chakras align with the rainbow of light that signifies the Visible Light Spectrum, wherein humanity exists. Even humanity's light field expresses separateness and difference. But when the seven chakras unite, their separate colours merge to form a channel of white light through the centre of your being.

To illustrate this more clearly, it might help to picture a stream of white light (visible light) entering a prism, and immediately dividing into a showy rainbow of separated colours. Upon hitting matter, the white light splits into its constituent wavelengths, revealing its colourful inner nature. And by reversing the process, the separated rainbow colours reunite to emerge once again as a stream of pure white light...bright and clear. The individual colours are still present, however they're now dancing together in unity. By analogy, as humanity overcomes emotions, it rises above the state of separateness and difference that's prevalent in the world. This is the natural state of autistics, as the high frequency stream of white light begins to settle within their spinal channel. At the same time, connecting with a field of brilliant light that co-exists with the VLS; a parallel reality where higher information sits naturally on the higher frequencies of perceived genius. That brilliant field of light is the imagination.

> *Somewhere over the rainbow, way up high…where happy little bluebirds fly …*

Harold Arlen and Yip Harburg, The Wizard of Oz (1940) Movie MGM.

Evolution is slowly but surely leading us *all* beyond the rainbow of the VLS, into the higher frequencies of clarity, brilliance and joy within the universal field of creation. Evolution is refining your emotions, your essence, your energies, as it opens your imagination. Evolution is inspiring you to harness your inner power, thus uniting all your chakra energies and creating a channel of pure white light through the centre of your being. This is the proverbial **tunnel**; from your base to your crown, in parallel to the evolving neuro-endocrine systems of the human race. This is where evolution is occurring. **[human & human race: see glossary]**

The macrocosmic notion of the *human race* defines a vast collective of all the peoples of the world. However, from a microcosmic perspective, the human *race* sits within each and every individual human being, from the base chakra to the crown chakra, indicating the measure of your unique mental, emotional and physical qualities.

The human *race* is the *channel* through which your nitrogenous humours inform every tissue of your mortal human body, through the central nervous system.

Evolution is transforming the central channel of energy; transforming the *human* race, by refining humanity's beliefs, behaviours and body tissue; slowing the demise of body organs and the leaking of energy; thus, reinstating the inner power that lights up the channel with a higher collective vision. Ease and youth are the rewards for changing your beliefs and overcoming your emotions, as you open to a higher vision for yourself and humanity.

I'd like you to imagine, from this new and higher perspective, just how beautiful your body would feel. And now, I'd like you to imagine how your body would feel if you were suddenly thrown back into the dense and smelly emotional blanket of the VLS with all its emotional baggage. Those exquisite sensitivities of your highly refined body would feel foreign, like a stranger in a parallel universe. This is how autistics feel *all the time*. Their information system is trying to establish its new channel of light while living within the imposing forces of this dense electromagnetic world. An unenviable task.

This is the evolutionary battle that's being played out in the limbic system, where you establish your state of mind...and matter.

The word *mind* is not only defined as your mental faculty, but is also the practice of taking care of, looking after or tending to something. You mind your children and your precious belongings, as well as your p's and q's. You also mind your personal beliefs and tend to them regularly, as they build your life story and infuse your body tissue, thus creating your reality. The Oxford Dictionary of Etymology provides a very clear connection with the limbic system when it defines the word *mind* as memory.

However, the limbic system doesn't act alone, for the treasures it stores are derived from the fizz of electrical firings that create your thought patterns from within your nervous system. Hence the more common definition of *mind* as being thought or intention. The limbic system *minds* the programmed thoughts and beliefs that are created by your nervous system. It acts like a library that's looking after your favourite stories; or a nursery where you hear constant retellings of your favourite childhood tales. In fact, the limbic system is exactly like a nursery, for it stores all your childhood memories. And thus, the *mind* is born bearing the seeds of thought that are sewn and nourished during childhood, and which blossom into the walled garden of your limited, adult imagination.

From this place of minding, your limbic system informs your endocrine system, which determines the nature of your body matter, as discussed earlier. Therefore, your mind determines the form and function of your matter. Your mind and matter connect through the limbic system, clearly illustrating its direct relationship with your chakras, which provide the energetic blueprint of this meeting. Your chakras reflect the energies of your limbic system as they course through your organs, indicating the loss of power that's inherent in your stored memories; the energies you give away to the external world because of the childhood stories you hold onto and massage repetitively as your familiar beliefs...family favourites.

The colours of your life build your persona from habit; your individual identity; your mask of separateness or difference, which leaks into the electromagnetic field of visible light as the script for your life reality.

> The electro-magnetic field is an emotional blanket of separateness and difference; a masquerade of personas.

We're evolving beyond separateness and difference by refining our library of beliefs. This is not only reflected in our limbic system but also in the energies of our chakras, as well as the electromagnetic field.

So, by taking the evolutionary progression one step further, we find that it's soothing the fiery nervous critic, diminishing the store of fearful and emotional memories, promoting more optimal form and functioning within your body organs, and reducing your energy leaks (emotions) into the electromagnetic field, thus maintaining increased power within your body tissue, fostering increased health and beauty. When the causes of deterioration are removed, your body regains the creative inclination of its original design. Youth.

This is a big jump that definitely requires a leap of faith from your belief system. But isn't that what fosters evolution? It really is common sense, which I hope to convey throughout this chapter. But first, something else to challenge your preconceptions.

It stands to reason that if the nervous and limbic systems are becoming obsolete, then so too must the *mind* also be diminishing into nothingness. For the mind is our limited and limiting store of information, our own personal library of childhood stories that we've gathered from our unique experiences, which differ for each and every one of us. Only when we operate from the place of the mind, do we experience difference. If the goal of evolution is unity, then the mind is on its way out.

How is this possible?

In order for billions of different minds to exist, there must be an eternal source of collective information into which we each dip to gather our unique blend of knowledge and belief. We store this info in our little limbic memory safes (personal library), and pass it determinedly through our ancestral lines, so as not to lose the thread; this we call tradition. Unfortunately, because we cannot see the eternal spring, or know of its ever-present nature, we place all our energy into our *little* minds, and keep re-visiting the same old stories, because we're programmed to do so. We literally miss the big picture. We are simply store-houses, accommodating a small portion of the eternal spring, which we express as our unique energy of creation. This is our evolutionary duty. Thankfully, with the expansive habit of each new generation, humanity is connecting with more and more of the eternal spring and gathering a clearer vision of evolution's exquisite plan. The small mind of difference is stepping aside, and warmly welcoming the universal flow of creation as the big picture unfolds its powerful and all-encompassing vision. You're graduating from your personal little limbic library to the expansive universal library, with its eternal supply of possibilities; the fully open garden of imagination; the source of *all* creation. **[tradition: see glossary]**

So, yes, the human mind *is* becoming obsolete and making way for the fully open imagination. You're switching gears to a new operating system. Evolution is asking you to declutter your defensive limbic library; to clear out those old stories that no longer serve you or humanity, so you can optimally engage your imagination, and reconnect with the full spectrum of creativity within the eternal spring of knowledge. Autistics have already made the switch, however it's difficult to utilise the new system when living in a world that is still invested in the clunky old mechanical mind. A relaxing of the old ways is required, so a new vision can unfold.

From mind to imagination; from darkness into light; from mortality into youth. This is the evolution of the human body, which is clearly evident in increasing longevity throughout the ages.

How exactly is evolution achieving this phenomenon within the fibres of your being?

Let's step back into the limbic brain and follow its trail of scent through your pineal and into your endocrine system. Effectively, this trail of scent is your little aromatic quantum of imagination expressing through your physical form. Each time you re-imagine your beliefs into higher possibilities and update your personal story, you not only forgive your tired old memories, but also purify your chemistry, sending increasingly beautiful bouquets of scent into your endocrine organs.

Your little package of imagination is therefore delivered by your ever-changing hormones, carrying the specific olfactory notes of your family's current chemical formula. Stretching your imagination evolves these genetic blossoms of information by refining the aromatic signature of your chromosomes, and transforming your body into a garden of delights, as evidenced by the innovative expressions of autistics.

It's your outdated hormonal family secrets that hinder your imagination and limit the free expression of your true essence through your endocrine organs and body tissue. Each endocrine organ has a different form and function within the human body, and therefore requires specific hormonal information. Each endocrine organ stores different family secrets; different bundles of emotional chemistry; different bouquets of scent. Each endocrine organ has its unique personality as it goes about its duties, reacting and responding according to its prescribed family formula.

And it's these individual endocrine personalities that are expressed through your chakras, appropriately attired in the different colours of the rainbow, as they adorn your body's spinal channel. Seven little packages of childhood programming that have been designed by your family patterns of belief and behaviour. Seven little childlike bundles living together in communion within your physical home, creating the overall condition of your body temple.

This is beginning to sound like the fairytale I promised you. Seven childlike bundles uniting through a pure white influence; Snow White and the Seven Dwarfs. In the next chapter you'll see that the analogy is indisputable, aligning with both autism and evolution.

I'll also be introducing the skeleton as an unexpected endocrine chamber. And as the emblem of death, it will avail itself to the seven dwarfs in the figurative form of Snow White's coffin.

Chapter Eight: Snow White and The Skeleton

Skeleton: f Latin *sceleton* the bony framework of the body, dried-up, mummy, parch, wither...
supporting frame...
bare or mere outline, rough draft.

Online Etymology Dictionary

Accordingly, your skeleton is the dried-up framework that provides the bare bones upon which your life story is fleshed out.

Not a glowing description by any means. So, what *is* it that we're missing in our understanding of the skeleton, that leads us to blindly accept its withering nature as normal?

That's exactly what Snow White will bring to light in this chapter, with the help of her seven little bundles of diverse energy. Let's step into the dark woods and excavate the truth about the skeleton; the truth that's already revealing itself in autistics.

I'm going to begin with those proverbial *skeletons in the closet* that represent the source of secret shame to a family, and thus literally provide the skeleton with its ever-diminishing evolutionary role as a vessel of disgrace that's fed by your family through your endocrine system. **[endocrine: see glossary]**

For, as we've discussed in previous chapters, the endocrine carries your in-house, or family, secrets; the chemical secretions that derive from your programmed fears; the memories and emotions that must be kept in the dark for fear of disapproval from the external world; your unique vulnerabilities in hiding, ready to react at the slightest provocation.

When deposited within the skeletal chamber, these family patterns of religious exactitude darken its cells with a heavy emotional burden, prompting its demise into bone.

Yes, I believe that bone is the deteriorated state of your skeleton.

And so it is that the dutiful skeleton is fashioned into the emblem of death. Mortality, it would seem, is a habit.

> *Rigid, the skeleton of habit alone upholds the human frame.*
>
> *Virginia Woolf (1925), Mrs. Dalloway.*

The *human* frame, in its bony *mortal* form, wears the calcified, rigid expression of your familiar habits. Your family's disgrace is mortifying to your skeleton. **[mortal: see glossary]**

mortal = mortar + -al = pertaining to mortar

Interestingly, bone bears an uncanny resemblance to mortar, which is the thick hardening paste that's used in the building industry to bond bricks or stone, ensuring a rigid, inflexible structure. It consists of lime, sand and water, the exact components of the rigid, inflexible ossified or bony skeleton...calcium (lime), silica (sand) and water.

> *Within bone, silica is the essential component making up the collagen matrix upon which calcium is deposited. This relationship is so*

> *fundamental that it is truly impossible to form bone without both calcium and silica.*

Eidon Iconic Minerals. Silica: A Little-Known Element Comes of Age.

The word *deposit* literally means to *dethrone* or remove from power; to depose, uncrown, usurp, dislodge, overthrow.

In the formation of bone, calcium-deposits overthrow the original silica to create an inferior collagen matrix that's devoid of power. Calcium de-thrones collagen.

> *Gradual calcium deposition within collagen occurs as a natural function of aging.*
>
> *Collagen, Wikipedia*

Calcium deposition confers degradation upon the collagen matrix, as it turns your skeleton into bone. Collagen no longer holds power; usurped by the family, who bestows the mortal crown upon your human frame. As you lose power, your lights dim, and your skeleton begins to wither and age.

Matrix is from the same derivative as *mother*. The degradation of your collagen matrix contributes to the decline and aging of your skeleton, as its *mother*-nature withers into bone.

Petrified of rocking the family boat, your *skeletons in the closet* form a bony coffin around which your life story clings in fleshy threads of fear.

However, prior to fear and degradation, in its youthful state, your skeleton is a chamber of light; a lighthouse wherein your imagination and sparkling innovation house your wishes and dreams, the creative expressions of your pineal. As previously revealed, the radiant pineal is the light at the end of your spinal tunnel; the flourishing garden of your imagination that eloquently advocates your uniqueness through the platform of your brilliant skeleton.

This is the goal of evolution; to overcome all the family secrets; to release all their rules and limitations; to dissolve all their fearful programs; so as to reinstate your original power and pure *mother*-nature. Your skeleton is the arena within which this spectral climb ascends.

And it's doing so through the cleansing of your limbic and neuroendocrine systems, through the purifying of your personal beliefs and stories, and through the generational forgiveness of old habits. Your skeleton is being increasingly bathed in the light of grace, courtesy of your opening pineal, in the long-running baptism called evolution.

Your pineal and skeleton inspire each other. What transpires in the pineal, expresses through the skeleton. The adult's atrophied pineal, is echoed in their bony skeleton. No coincidence, as both *atrophy* and *skeleton* share the same meaning; withered and dried up. The imagination withers and dries up on its regimental march toward the closure of the adult pineal. Fortunately, with a little nourishment and inspiration, it's been slowly rejuvenating on its course through evolution; your third eye is opening in a blaze of enlightenment, which your skeleton flaunts through its increasingly radiant fibres. In Chapter Five, I revealed that the characteristics of autistics aligned with those of an open third eye. It stands to reason, therefore, that the autistic skeleton is communicating with their fully functioning pineal, and that innovation and imagination are lighting up its structure. Therefore, far from being dysfunctional, the autistic skeleton is, in fact, more highly evolved as its tissue embraces the higher light frequencies of a vivid imagination. The autistic skeleton is taking back its power, and reinstating its original telepathic communication system within its carbon fibres.

This is the original, embryonic infant state of your skeleton. And it's this shiny skeleton that, so long ago, made its presence felt as the radiance of your pregnant mother. Your brilliance was a given...and is a gift that still sits deep within you. This is your first and fondest memory; your guiding star that's calling you home; your skeleton in its snowy white form. **[infancy & infant: see glossary]**

Figuratively, Snow White resides in your skeleton in the Once Upon a Time infancy of your life fairytale, before the onset of the impacting external world and the formation of your bank of memories and emotions. She radiates the same gentle grace and kindness of spirit that's prevalent within autistics, and features in those with an open third eye. Autistics' imaginations are flourishing and soaring into the heights of those telepathic geese...and infants. Autistics, Snow White, geese and infants are neighbourhood acquaintances from the higher spectrum of light frequencies where innovation, fantasy, telepathy and joy already exist, awaiting your belief. They're waiting for you to remember your truth, so you can join them in celebrating your precious gift.

Snow White is going to help jog your memory.

As we revisit the Once Upon a Time of the fairy tale itself, we find Snow White to be the fairest in the land, sweet, kind and innocent. She's yet to harbour any fears or doubts, and so epitomises the first and fondest memory of your infancy, in open communication with your own personal wishing well, your pineal. It's no surprise that the pure energies of Snow White conjure princely dreams that echo her sentiments of pure love within a united heart. Your snowy white infant skeleton, which sits at the heart of your body, unites every fibre of your being through the pure energies of her nurturing, motherly love. *Mother*-nature at her finest. Your connective tissue sings in unison, for your infant spinal channel is yet to separate into its seven little secret endocrine bundles. The seven dwarfs are yet to appear. And Prince Charming is yet to ride in on his white charger. **[Prince Charming: see glossary]**

Prince Charming rescuing Snow White gets up the nose of many a female. But rest assured that this is simply an illusion that's stood the test of a patriarchal mindset. For, believe me, there's not one fairy tale wherein a *man* rescues a *woman*. Let's find that elusive Prince and put this fallacy to bed, so to speak, once and for all. It's time to open your imaginations to a new…and infinitely dreamier… fancy.

The word *prince* means first or prime, which extends to being the finest or choicest part, otherwise known as the flower. (Oxford Dictionary of Etymology) From this new understanding, the prince's *dreamy* fairy tale personification isn't an external offering that will bring happiness *to* you, but instead pertains to *your* most exquisite dreams and fondest yearnings; that which you hold in your own heart. As the choicest part or flower, the prince is analogous to the most beautiful essence or scent that expresses *your* most imaginative dreams into reality, when fertilised with belief.

Therefore, every time you re-imagine your dreams and beliefs into a higher essence, as you lovingly nurture them, you upgrade your imagination and conjure a more princely reality for yourself and your body tissue. Autistics are masters of the upgrade with their penchant for the nurturing of innovation. Autistics epitomise Prince Charming.

Whatever you imagine, you attract, for your imagination is magnetic. And magnetism is charm. Therefore, Prince Charming literally translates as *your* prime magnetism; *your* most attractive essence; *your* most beautiful inner garden in full flower; *your* flourishing imagination expressing through your body tissue as purity and youth. This imaginative and youthful magnetic essence blossoms within your body's energy meridians enhancing your power of attraction.

Fairy tales are asking *you* to rescue *yourself* by opening your energy channels to a higher creative force. Prince Charming is your own power. I've said it before, but I'll say it again, your imagination is the most powerful force in the universe. Dancing with Prince Charming simply means that you're powerfully attractive so all your dreams are coming true, because you're expressing your full imagination. Feed and water it with as much belief, and love, as you can. Your inner belief feeds the inner garden of your imagination, thus nurturing and growing your inner creative power...your unique creative genius.

Your imagination is always creative; never destructive. Consequently, your imagination does not support aging body tissue; it fosters youth. This is clearly illustrated in the youthful nature of the highly imaginative, snowy white infant skeleton. When you dance with prince charming, your creative outlook is rescuing your body from the ravages of TIME.

Rose-coloured glasses? Well, why not?

No surprises that the prince in the fairytale Snow White is called Prince *Florimond*; the prince of flowers. He represents the choicest floral essence...rose, which is the purest form expressing through matter, as witnessed in the soft rosy glow of an infant's complexion.

MATTER = E.M. ATTAR = perfume of roses

> *Rose has the highest frequency of any oil, and raises the frequency of cells, bringing harmony and well-being to the body and balancing personal will...offers protection by raising the energy field frequency.*
>
> *Rose symbolizes divine or mystical love...Rose activates feminine energies.*
>
> *Aromatica Medica*

Prince Charming is the blossom of your reality; the flowering of your dreams; your dreams come true. Prince Charming is your flourishing imagination. Prince Charming is *you* at your most magnetic, your body a beautifully scented garden, presenting the finest bouquet of roses. Prince Charming is *you* romancing yourself. When in this state, you'll find that everything in your life is coming up roses!

Prince Charming is also the happily ever after in the fairytale called evolution, which is fostering the inward shift toward genius. Your flourishing imagination *is* your genius. Autistics are clearly forerunners of evolution as their imaginations leap and prance into innovative spheres.

Only from within can evolution be re-imagined.

Meanwhile, back in the fairy tale, we find the threatening presence of the external world wielding its control over Snow White in the guise of the wicked queen's jealous fears. Her magic mirror represents the electro-magnetic field, reflecting expectation and accepted norms; celebrating appearances over inner beauty; emotions over feelings; the ego over the collective; the mask of the persona in full voice. The external world is an insecure construct of fear.

The electromagnetic field is essentially a mirror of light that reflects the emotional state of humanity, in an enormous aromatic feedback loop of fear. This is the domain of the wicked *step*-mother, for she is analogous to *mother*-nature in her saddened form; an unhappy outer reality; aging matter. She is not Snow White's *real* mother, for the true design of your *mother*-nature can only be expressed through nurturing love from within.

As the fairytale plays out within your childhood, the established egoic traits of a wicked world begin to mask the goodness and purity of your true innocence and beauty. However, your inner protector (the kindly woodsman = immune system) will never turn on your innocent nature, opting instead to secrete it in the dark depths of the worldly woods. Unfortunately, although protected, your beautiful spirit of light gets lost in the shadows of mediocrity and tradition, memory and emotion. The lights begin to dim in your snowy white skeleton. Your true essence is camouflaged behind a mask of hormones in order to survive the perceived threats from the worldly woods. **[woods: see glossary]**

> *...in our childhood, we might have laid down our own traumatic memories in our bones.*
>
> *Sue Black (2020), Written In Bone: hidden stories in what we leave behind (page 232)*

Hormones are impulses that regulate your activity and behaviour; instincts that incite automatic reactions. Interestingly, the word instinct derives from *in + stinguere* meaning *in + prick*. Hormones are the cyclical, repetitive weavers of the emotional electro-magnetic tapestry that is your persona, behind which your pure, innocent beauty slumbers. The instinctive prick of hormones upon the spinning wheel of life invokes strains of another fairy tale. **[instinct: see glossary]**

Instinct suppresses intuition. Instinct and intuition are very different forces. Instinct is indicative of your wicked stepmother, while intuition is your original *mother*-nature. Instinct is fed by fear and is externally driven. Intuition is the loving guidance of your first and fondest memory expressing from within. It's your *instinctive* hormones that drive your little light into hiding; your family shutting down your individual voice; your smelly chemistry masking your pure rosy essence;

destruction replacing creation; age replacing youth and beauty. The smelly prisoners of your instinctive programmed behaviours cower in your dark cells defining your persona...and your bony skeleton. **[intuition: see glossary]**

You'll find the seven dwarfs inhabiting your cells, for their unique individual characteristics represent the emotional habits that define your persona...and your physical habitat, where they dwell awaiting rehabilitation through the process of evolution.

Snow White's fairy tale journey into the darkness leads her to the cottage of the seven dwarfs, just as your sojourn into the dense worldly woods of life leads to your dark cells. The dwarf's cottage is indicative of your cells, with the seven dwarfs representing the colourful array of traits that reside therein; the chemical reactions that define your emotions; the hormones carrying your signature family patterns; your aromatic fingerprint; the sevenfold rainbow of the lower light field within the visible light spectrum.

Only when Snow White enters the dark woods does she feel fear...in her bones. Instinct threatens to snuff out intuition.

Accordingly, your cells darken as your infant light dims through childhood programming from the rigid external world. Your bright light of genius fades as your imagination is pruned and shaped into the shrunken family bible of limitations. Cobwebs gather, indicating time, aging and memories...evidence of an active limbic system...your defences are up. And mothering love slips into the shadows as fear sets in; your *nervous* system is in need of soothing. Your inner power is switched off.

The seven dwarfs and their cottage personify all the shameful family secrets within the cells of your endocrine organs, including your skeleton; and the formula of chemical reactivity that deteriorates your body tissue, creating aging and mortality. The family secrets (emotions) go to work in your skeleton, just like the seven dwarfs dig, dig, digging in their dark cave;

excavating coal, the fossilised remnants of that which has been buried and transformed through heat and pressure. Through the heat and pressure of childhood, your family eats away your shiny white skeleton as it settles into the re-fashioned bony framework upon which you flesh out your life story via a script of emotion. **[bone: see glossary]**

Clearly the dutiful seven dwarfs are miners, doing the job they're programmed to do with habitual regularity. Just as children (minors) do what *they're* programmed to do through habitual family reinforcement. It's your childhood programming that's eating away at your skeleton, depositing the inorganic mineral content, its earthly nature, within your bones. **[genome: see glossary]**

To mineralize means to:

> *...convert (organic matter) wholly or partly into a mineral or inorganic material or structure.*
>
> *Online Oxford Languages Dictionary*

Interestingly, to mineralize also means to petrify! To hold so much fear that you're unable to move or feel; to be terrified; to be scared stiff! **[terra firma: see glossary]**

> scared stiff = terror firmer = terra firma = earth

The organic matter of your original snowy white infant skeleton gradually mineralizes as the child treads carefully through the minefield of external experiences, mapping the family boundaries in the limbic atlas.

A system of fear is being encoded into their body tissue; they don't feel safe...*the* **nervous** *system*. And each experience is catalogued for future reference; they remember the pain...*the* *limbic system*. Then chemical soldiers are deployed to attack the perceived threat; they react emotionally... *the* **endocrine** *system*.

The issue is in the tissue.

This battleground of fear is the child's own body tissue. Fear is eating away their integrity, both figuratively and literally. Unsuspectingly, their true feelings have been petrified into submission, and so they're incapable of doing anything about it. Within their connective tissue, family custom is being set in stone, inflexible and intimidating. **[connect & connective tissue: see glossary** (necessary for all readers please)**]**

At this point I feel it's important to look at the very thing that makes your existence possible; that carries your essence, your life story; that forms your physical body. This is your connective tissue, which is carbon-based and makes up your body matter. It has a thankless task, and quite frankly flies under the radar, unless it presents with an unfavourable condition, at which time it draws your attention. **[carbon: see glossary]**

Your connective tissue is literally that which connects you magically to everything and everyone, via the energetic frequency pathways that create the ever-present subtle, intuitive, universal communication system. Awake or asleep, you're constantly switched on and surfing these wireless communication highways; your connective tissue sending and receiving energetic messages on various light frequencies.

Higher frequencies use less energy, giving away less power. Accordingly, your body tissue doesn't need to work so hard, therefore reducing its levels of wear and tear. This is why evolution is inviting you to refine your beliefs and behaviours, so you too can treat your connective tissue with kindness, and reinstate its ease and youth.

And, as an aside, because you use less power on the higher frequencies, your connective tissue requires less recharging of its batteries (cells), in other words, less sleep or rest. All those perceived sleep problems within the autistic tribe, suddenly take on a new and far more palatable explanation. Autistics are forerunners of evolution, with higher frequencies running through their connective tissue. Can I remind you once again of those high-flying geese!

When aligned with the glossary entry, your connective tissue is *tissue that overcomes death*. Although this probably seems outrageous, it's actually common sense, because where else in the body would the evolutionary refinement take place if not in your very tissue?

The aromatic fingerprint of all matter (DNA) is based within the flower garden of your carbon fibres, forming the connective tissue that carries your essence, whatever its measure of sweetness.

Evolution provides you with this opportunity to refine your sweet essence by progressively overcoming the limitations inherent in your DNA blue-print, and ultimately infusing your skeleton with the sweetest essence of roses. In fairy tale parlance, Snow White dances with Prince Charming, as you connect with your dreams...all within your own body.

By overcoming fear, you *come-up-over-fear* into the higher, sweeter light frequencies of love. By overcoming petrification, mineralisation, your skeleton comes-up-over its bony condition into the cartilaginous form of youth.

Another meaning for petrified is *ossified*...to become rigid or fixed in attitude. The petrified skeleton becomes rigid and stony through the process of ossification. And this is considered normal?

The withering nature of your skeleton is considered normal in this mortal world where aging and deterioration are expected, even anticipated. For this is what science sees on a regular basis. There's no room for imagination in science. Science deals with facts...also rigid and fixed.

As a result, when the autistic skeleton presents structural facts that don't fit the withering norm, it's immediately relegated to being disordered and dysfunctional...it's different...it doesn't fit the facts! **[fact: see glossary]**

> *Never give up on what you really want to do. The person with big dreams is more powerful than the one with all the facts.*
>
> *Albert Einstein*

Surely the facts speak for themselves. Science is bound; limited to what's already known. Science is the knowledge that's already contained within your limbic vault; your little, earthly mind. Science is memory. Science creates mortality. Evolution is very grateful to science for recording the history of knowledge, bit by bit, within your brain, but it's time to release the brakes and open the flood waters of the imagination, so as to liberate both your brain...and your body. The dam has already burst for Autistics, allowing them to tap into the eternal spring of knowledge, thus providing the opportunity for them to dive in and retrieve whatever information they require at any time they desire, particularly in their areas of interest. Science and imagination have aligned for them.

Neurotypicals find these tendencies quite magical, and have even coined the term savant to define them. However, this too is a misnomer, for *savant* means *a man of learning*, from the same Latin derivative as the word *science*. Autistics do not need to learn and retain information in their little scientific limbic brains in the way that neurotypicals do, as I've already explained. Imaginative or innovative or creative or magical or wondrous are far better definitions of the autistic gift.

But it's not just an *autistic* gift! It's the gift that evolution is bestowing upon everyone. It's the gift that dearly wishes to flow freely through your connective tissue and light up your skeleton.

I asked the question at the beginning of this chapter, "what is it that we're missing in the understanding of the skeleton, that leads us to blindly accept it's withering nature as normal?" Clearly, it's the imagination that's missing. Your skeleton turns to bone because its tissue ('tis you) is missing your spirited imagination. It's petrified without it. *'Tis **you*** that's missing your own spirit; your true essence; your wondrous imagination that defines your uniqueness (*you*niqueness). Your beautiful little light is missing from your physical home, and without it, you're petrified of the dark. Your skeletal connective tissue comprises the threads of your life; the yarns that you've collected; the fairy tale that narrates your character, and it's missing its authenticity...you. The authentic *you* is biding her time in the eternal spring awaiting your re-imagined reality.

What we *think* is real...simply isn't. But we're so hypnotised by what we *see*, that we've lost the ability to imagine a *different* reality; one that is true to our heart. We're oblivious to alternative possibilities because we find comfort in what we already know, and fear the unknown. We literally fear our true self because it's so unfamiliar. We're blinded by habit. We're blinkered by mortality. I understand how science would find it hard to believe that the skeleton was a chamber of light holding eternal youth in its true form, simply because they haven't seen an example of it for themselves. But if they opened their imaginations to this *possibility*, they'd soon begin seeing the signs of its truth. And that goes for everyone. When you open your minds to the wonders of yourself, and push the boundaries into innovation, knowing that your body tissue is the canvas upon which your imagined reality is painted, you too will begin to physically see the signs. Your dreams will come true as you create magic in your very own connective tissue. But you have to believe.

And if you don't believe *me*, look to autistics for evidence.

Why is it that autistics have a low bone mineral density (BMD)? Their skeletons are less mineralized, less ossified, less petrified, less mortified. The dig, dig, digging of the seven dwarfs isn't as prevalent in the autistic skeleton. Their skeletal closets hold fewer family secrets. The chemical reactivity of the emotional persona is reduced, along with the fear factor that drives it, making sense of their perceived under-functioning limbic system.

I've previously explained that the limbic system is where the evolution of family patterns is recorded. The shrunken family catalogue of the limbic information passes through the endocrine, prompting the skeleton's deterioration into bone. The unimaginative limbic information creates bone. It stands to reason, therefore, that evolution is not only rendering the limbic system obsolete, but also the bony nature of the skeleton. Science will begin to encounter an increased prevalence of low bone mineral density in alignment with the forward movement of evolution and its flourishing imagination. The autistic skeleton is already providing such evidence. The autistic skeleton bodes well for a future of innovation, imagination and creative genius...how brilliant!

Imagination is more important than knowledge.

Knowledge is limited.

Imagination encircles the world.

Albert Einstein

Within the limbic system, family patterns of *emotion* and *smell* are stored as one memory. Therefore, science takes the form of smell. This scientific information (your chemistry/DNA), your aromatic fingerprint, is passed through the pineal, eventually alighting in the skeleton. Smell forms the grapevine of communication throughout the body. The skeleton stores family patterns in the form of smell within its connective tissue.

However, as the family story evolves within the limbic system, smell evolves within the skeleton. Evolution is recorded within the structure of the skeleton. And if you look closely, you'll even find essential oils tucked within the word *evolution*.
[evolution 3: see glossary]

Autistics' limbic systems are clearly more highly evolved; their personas are more relaxed; their skeletons are less dense because they carry a more purified essence and radiant light. The seven dwarfs have remained lovingly united within the autistic skeleton, for they intuitively maintain their gracious connection to their snowy white infancy.

In the fairy tale, the seven dwarfs are seen to rehabilitate their unique individual habits in the presence of Snow White. They're inspired to re-imagine themselves as they bask in the glow of her beautiful energy. Snow White cleanses their dwelling, just as the nurturing light of your own *mother*-nature cleanses your cells, lifting your skeletal hearts into higher frequencies.

Your dense electromagnetic neuro-endocrine system will bid farewell and take flight into higher energies, having played its part in the ascendency of your spirited imagination.

Grace is being reinstated as the true essence of your skeleton. In the long-running baptism of evolution, the sins of the past are dissolving with the onset of an individual and collective forgiveness. That's Snow White's gift to the seven dwarfs. And that's the gift you bestow upon your cells when your hearts are high. Through forgiveness, the heart of your every cell releases the unique aromatic fingerprint of your family's DNA, opening you to a more fully enlightened and universal imagination. And the emblematic death bed of your skeleton transforms into a bed of roses. This cleansing process forgives all the family secrets, traditions, patterns and hurts; all the smelly prisoners that have been wafting through your ancestral line throughout evolution. Your own rosy outlook is watering the garden of your imagination, nurturing, mothering and liberating it from its dried-up skeletal coffin. **[sarcophagus: see glossary]**

I simply must address the issue of Snow White's coffin that I promised at the end of the previous chapter. The reputed flesh-eating calcium-based limestone, which gives the sarcophagus its name, is indicative of your bony skeletal coffin after its youthful snowy white nature has been mineralised/petrified/ossified. Interestingly, silica is used to make glass, so the calcium deposits that denature the silica in your skeleton, send both it and your snowy white nature to their death. Snow White's glass coffin is your denatured skeleton.

That said, let's now look at how your rosy outlook is watering the garden of your imagination and liberating it from its dried-up skeletal coffin.

The feminine always equates to water, just as the masculine equates to fire. So, it's not surprising that your nurturing, motherly feminine energies provide the rejuvenating waters for a skeleton that's thirsty for love.

> *water: transparent liquid forming the material of seas, lakes and rivers;*
>
> *(prob. after Arab ma' water, lustre, splendour) transparency and lustre of a gem.*
>
> *Oxford English Dictionary of Etymology*

Forgiveness, baptism and cleansing usually involve the presence of water. Interestingly, as seen in the above definition, water has two distinct meanings. Most commonly, its liquid form, which is so crucial to your survival; and surprisingly, the splendour, transparency and lustre of a gem. In other words, its clarity and light.

In the study of gems, the outer appearance is measured by the rainbow of twinkling lights that dance on its surface, which is called its *fire*. But it's true worth is nestled deep at the heart of the gem, and is called its *water*, which is the measure of its purity and truth; its clarity and brilliance.

Both literally and figuratively, when something is said to be *of the highest water*, it's of the highest quality, value or worth. The seven dwarfs, with their electromagnetic rainbow of emotions, are indicative of the fire dancing on the surface of your bony skeleton. And Snow White represents the purity and clarity of your light-filled skeleton; your body-temple lighting up from within as your spirit arrives home after its long evolutionary trek through the dark woods, having rediscovered its brilliance...it's genius...it's crystal-clear heart.

A light and pure heart holds the highest worth; when you remember just how brilliant you really are, and how wondrous your body can be if you cherish it and fill it with love.

In Chapter Six, I used Pinocchio and Geppetto to reveal the monstrous lie of mortality; that we believe we're mortal beings of so little worth that we have no power over our own bodies; that we're programmed to believe we're not good enough to sustain the loving caress of our original *mother*-nature, our first and fondest memory, within the fibres of our being; that we're undeserving of youth and beauty, our snowy-white purity. Evolution forgives this lie every time you bathe your cells in the light of truth from deep within your own loving heart; every time you romance yourself. You *are* worthy of feeling this love. You *are* worthy of feeling good. Your rosy outlook does indeed water the garden of your imagination with its clarity and brilliance, reinstating integrity within your connective tissue...and overcoming the mortal lie. **[truth: see glossary]**

> *Almost every area of our body, soft tissue and hard, carry an echo of our experiences, our habits and our activities...Many of these memories remain locked within our skeleton.*
>
> *Sue Black (2020), Written In Bone: hidden stories in what we leave behind*

Your skeleton holds your story...

Chapter Nine: Story

The stories that our bodies narrate cannot be divorced from the conditions of our existence.

Nancy Krieger (American epidemiologist), cited in Written In Bone by Sue Black, p 327

Story is everything.

And everything is story.

Story stands shoulder to shoulder with reality.

> *Any large-scale human cooperation...is rooted in common myths that exist only in people's collective imagination...revolved around telling stories and convincing people to believe them. (p 30-1)*

and this...

> *The immense diversity of imagined realities that Sapiens invented and the resulting diversity of behaviour patterns, are the main components of what we call cultures. Once cultures appeared, they never ceased to change and develop, and these unstoppable alterations are what we call 'history'. (p 40)*

and this...

> *Thanks to their ability to invent fiction, Sapiens create more and more complex games, which each generation develops and elaborates even further. (p 43)*

Yuval Noah Harari (2011), Sapiens: a brief history of humankind (pages 30-31, 40, 43)

History, therefore, is a chain of stories flowing through the long-running narrative of evolution.

Hence, evolution is a fiction, that's dependent upon human belief to ground it in reality. It would seem that...

Reality is imagined

And so, the stories our bodies narrate from generation to generation, depict the imagined reality of the world around us, which in turn feeds back upon us, in a cyclical evolutionary challenge of ongoing boundary-pushing.

Likewise, the narrative expressing through the body tissue of autistics is reflective of the conditions of *their* existence; those conditions being two-fold and contradictory. Their intuitive evolutionary narrative, with its innovative nature and gentle inner voice, sits alongside the narrative of misunderstanding that's bestowed upon them by an external neurotypical world that has its own imagined reality...its own story...about autism.

What's considered to be the medical model of autism is, in fact, the story that's been exerting force upon the world, most particularly upon autistics and their families. This story, which is far from the truth, has been believed by the majority simply because it has a medical source, giving it the power to become reality. Unfortunately, once a story settles into the belief system of human beings, it's hard to shift...it takes time...it takes a new imagining.

It takes evolution.

However, if reality is imagined, then *you* possess enormous power. Very simply, you can create for yourself any life reality that you can imagine.

Let's now step into your body tissue and follow the thread of imaginings that's literally creating the storyline of evolution, as well as ultimately liberating autistics.

Written upon the parchment of your body tissue is what I believe to be the most beautiful fairy tale of all; lovingly whispered from generation to generation through your *mother* line; her maternal storyteller being your mitochondria.

Mitochondria are the powerhouses of your cells.

> *They generate most of the chemical energy needed to power the cells' biochemical reactions.*

Mitochondria: National Human Genome Research Institute (updated 2024)

If it's chemical, it's smell, so your cells' reactions are powered by smell. And by association, your cells' reactions are powered by memory. The mitochondrial story is the refinement of smell; the updating of ancestral memories. Your *mother*-line carries a powerful, essential evolutionary message in her storytelling.

Interestingly, after fertilisation, the oocyte, or egg, strips the sperm of *its* mitochondrial DNA.

> *This process...ensures exclusively maternal inheritance of the ... mitochondrial DNA genome...[which] is essential for normal*

fertilization and embryonic and fetal development.

Dalen Zuidema et al (2023), Identification of candidate mitochondrial inheritance determinants using the mammalian cell-free system.

Your body's powerhouses are maternal!
All power to the mother.

Your mitochondrial powerhouses are the generators, or creators, of energy. Your mother-line has a creative essence. And she's intent on contributing to the forward thrust of evolution by encouraging and fostering the imaginations of successive generations. **[generator: see glossary]**

Stripping the sperm's mDNA is part of her cunning plan. For in so doing, she dilutes the patriarchal story that's been foiling her creative efforts. By infusing it with increasing doses of her gentle and loving feminine essence, she's progressively magnifying her maternal qualities within humanity.

The patriarchal story is masculine by nature, and fire is the masculine energy force. Fire is indicative of anger, which takes the form of inflammation within the body tissue. Throughout evolution, your gentle, motherly storyteller is pacifying the fiery narrative in your cells.

The evolving maternal whisperings of your mitochondria are responsible for the rise in feminism and the increasing female presence in more recent history. Your *mother*-nature is finding her place in the evolutionary story, both in the world reality and within the fibres of your being.

She's your very own Mother Goose, passing nursery stories from mother to child down through the generations. Imbued with purity, innocence and love, they embody romance and magic and possibility. Your mitochondrial Mother Goose delivers the spirit of make-believe into your cells; she wants you to believe that anything's possible; she is the source of your imagination and dreams come true.

Your mitochondrial Mother Goose has a direct link, therefore, with your pineal, smell and skeleton, and an intimate association with your diminishing limbic system. Through her evolutionary cleansing, her nurturing influence is soothing your limbic fears and asking you to remember your true, loving *mother*-nature and feel her presence in your bones. **[mitochondria 1, 2 & 3: see glossary]**

The generational revisions to your aromatic fingerprint (DNA) ride through evolution on the back of your mitochondrial Mother Goose, gently activating your pineal and expressing through your skeleton. She recognises that her stories must be taken with a grain of salt, due to their patriarchal overtones, until you are ready to see them in a new maternal light, through your open third eye.

Let's start by opening your eyes to a vastly different perspective about the patriarchal baton of evolution, the Y chromosome. I ask you to remember that each and every individual, whether male or female, has a balance of masculine and feminine energies within their nature. Each are touched by the maternal and the paternal to some extent. However, it's this mix that is changing within the evolutionary narrative of the human body as the feminine releases her wounded past, thus casting off her temporary masculine mask of protection.

What part does the Y chromosome play in the unfoldment of the fairy tale of evolution?

Down through the years, as technology evolved, microscopes (and a happy accident) eventually provided scientists with the ability to clearly see 46 chromosomes within a human cell, forming the 23 pairs that we're now familiar with. Prior to this discovery, the exact number was unclear as the chromosomes bunched together in the petri dish. The new information also showed that there was always one chromosome that appeared to be unpaired, existing on its own. Scientists labelled this single unknown chromosome, X. However, they soon discovered that it did indeed have a partner in the genetic dance, albeit a much smaller version, which they labelled Y.

> *...dissimilar in size and band pattern from any other on the slide. The smaller of these sits at the edge of the field, slightly adrift from the others.*
>
> *It has no partner, nothing with which it can be matched.*
>
> *This is my Y-chromosome, the bearer of my maleness and the token passed unaltered down from a long line of fathers.*

Bryan Sykes (2003), *Adam's Curse: A Future Without Men* (page 46)

What I find bewildering is the fact that scientists found no cause for alarm in the peculiar appearance of this tiny chromosome, bearing no resemblance to any other chromosomes on the slide. And that, without question, it so casually continued its role as patriarchal successor in the ongoing narrative of sexual reproduction.

In *my* mind, far from being the hero, the Y chromosome is, in fact, the very mutation that caused the need for an evolutionary progression of refinement, and triggered round after round of *reproductions* that will eventually lead us back to *the original*. **[mutation: see glossary]**

The WHY? chromosome doesn't know the answers. And yet he's been the head of the family throughout history [HIS STORY].

The WHY? chromosome is not yet connected to the eternal

spring of universal knowledge; he believes in the facts that science currently presents; he sits, with science, in the limbic system of limitation, vehemently defending his ignorance, his world, his kingdom, and his family secrets. He's afraid of innovation because that is where he'll meet his demise. He's rigid, competitive, angry, emotional, sexual, controlling and full of fear. He wears the mask of the fiery reptilian, the masculine. But, in reality, he's simply the wounded feminine; the *mother*-nature who's taken up arms to protect her ailing flesh from the onslaught of the external world.

> *It is no secret that, underneath it all, men are basically genetically modified women.*

Bryan Sykes (2003), Adam's Curse: A Future Without Men (page 14)

The WHY? chromosome or wounded feminine has lost her power; she's feeling ill-at-ease, which manifests within her tissue as dis-ease; disease. She's a shadow of her former self in this mutated, shrivelled form, which is why she can only create *reproductions*, poor copies, instead of her original *mother*-nature, as she passes dis-ease from generation to generation. What a sad little mutation she is. What an angry little mutation she is.

> *mitochondria are known as the guardians of the inflammatory response*

Das Chatterjee et al. (2003), Mitochondrial Epigenetics Regulating Inflammation in Cancer and Aging. National Institute of Health.

When she senses that her body tissue is being attacked, your mitochondria immediately go into protective mode, employing inflammation as her first line of defence. Your *mother*-nature is angry because her charge is being threatened. She's now exhibiting the masculine behaviour of your chemical soldiers that exhibit her wounding in their fiery onslaught, bringing deterioration and aging to your body tissue.

Effectively, your fiery masculine takes on the personification of Lucifer, the fallen angel, as your cells become an angry inferno of inflammation; an abyss of disease. **[Lucifer: see glossary]**

The Y chromosome provides us with the reason for evolution; to refine the angry, rigid and limiting patriarchal expression of dis-ease that creates the *mortal* condition of MANkind; to eliminate inherited family disease. The feminine, creative essence of your mitochondrial DNA has been slowly but surely softening the patriarchal agenda, and stretching the human imagination to its innovative fullness. This is *her* evolutionary agenda; to infuse your connective tissue with her highest expression; united in love; bringing back youth, beauty and the fullness of health; and ultimately immortality.

Science calls this process of changes within gene expression, **epigenetics**.

> *Epigenetics is the study of how your behaviours and environment can cause changes that affect the way your genes work...epigenetic changes are reversible and...can change how your body reads a DNA sequence.*

> What Is Epigenetics? (2022), *Centres for Disease Control and Prevention: Genomics & Precision Health.*

This article also explains that,

> *Gene expression refers to how often or when proteins are created from the instructions within your genes.*

Proteins are simply habits, and therefore your gene expression determines the frequency and timing of your learned family habits. Your gene expression defines your beliefs; your story, which in turn determines how your cells function and take form. Your gene expression formulates your cells' chemical script; your smell; your memory. Tradition is inherent in proteins. So, epigenetic changes in gene expression evolve your aromatic fingerprint and consequently your physical form, as they update family tradition. **[generation 1 & 2: see glossary] [tradition: see glossary]**

Mitochondria are essential players in this epigenetic unfoldment of rehabilitation or refinement from patriarchal to feminine.

Having softened, the patriarchal agenda will have run its course, resuming its original feminine *mother*-nature; that which it had simply forgotten during its evolutionary slumber. **[limbo: see glossary]**

Clearly, everything in its true form is *mother* nature.

Limbo, as a place of confinement and forgetfulness, finds its counterpart in the brain's Limbic system, where our memories and emotions are confined, causing us to forget our true, innocent essence and *mother*-nature. This is where the patriarchal story sits throughout the evolutionary refinement into love. This is where mitochondria are freeing Lucifer from his fiery bondage.

Let me explain. You don't have to understand the chemistry, just come along for the ride as I create a simple analogy. Mitochondria create ATP (adenosine triphosphate), which is the energy we use and store in our cells. Within ATP, phosphorus is bonded to oxygen. So, phosphorus is essential to the energy in our cells. Now, here's where the analogy comes in. Phosphorus is another name for Lucifer, both meaning *light-bringer*. Lucifer, the fallen angel, is quite clearly indicative of the light-bringing, angelic infant who falls from the heavenly heights of the highest light frequencies through its childhood journey into adulthood, gathering emotional dis-ease in its fiery cells. By analogy, it's the radiant light-bringing properties of phosphorus within the infant that diminish through childhood and into adulthood. It's my belief that the soothing mitochondrial whisperings are massaging the oxygen bonds and loosening their hold on phosphorus, thus releasing the angelic light into your body tissue so it can resume its original form.

Arriving finally back at the original, the evolutionary chain of fathers (Y chromosomes), having served its purpose, will be obsolete, along with its cyclical reproductions. Bryan Sykes, in his book Adam's Curse tells us that the Y chromosome is, in fact, decaying at an alarming rate, and that,

> *...Like many species before us who have lost their males, we run the real risk of extinction. (page 16)*

Let me assure you that there's nothing to fear. From a big picture perspective, this is not a bleak prophesy, for it indicates that the evolutionary plan is indeed working. As she soothes her wounded Y chromosomal form, your *mother*-nature is progressively throwing off her reptilian, masculine mask of dis-ease; the Y chromosome is therefore diminishing in number, along with its inherent diseases, in alignment with the great plan of evolution. This is cause for celebration.

Being the curious soul that I am, I asked myself why it is that the Y chromosome hasn't altered on its journey through the evolutionary chain of fathers, being that the very nature of evolution is change or alteration. It became clear to me that the Y chromosome is the anchor of tradition; that her woundedness needs to hold onto long established custom; that her fear requires the familiar in order to feel secure. The familiar takes the form of family patterns of belief and behaviour that are inherent in hormones; the chemical soldiers that perform smelly patriarchal assignments within the cells in an attempt to maintain control. Tradition likes to reproduce itself; make identical copies that pass from generation to generation, in an attempt to stay the same. Tradition is a security blanket, which the Y chromosome is clutching onto *because* she's been wounded in the past and is afraid of being hurt again. Tradition, along with her fearful impetus, is inherent in sexual reproduction; in your DNA; in your genes, anchoring your emotional dis-ease into your protein structure. Effectively, the Y chromosome *is* altering, albeit with cunning maternal subtlety. As she evolves her wounded masculine traits, their presence diminishes, along with tradition. So, the cycles of sexual reproduction will disappear with the transformation of the Y chromosome back to her original feminine *mother*-nature.

Accordingly, the sexes will also disappear. Are we not already seeing signs of this phenomenon as gender identity-trends rattle accepted norms?

SEX = GENDER = AGENDA = It's why we're here.

Gender has given men and women their respective agendas throughout history. These specific male and female characteristics become apparent within your body tissue, being the arena through which the patriarchal and mitochondrial agendas play out the dance of evolution. **[gender: see glossary]**

The characteristics ascribed to each *gender* are accompanied by the hidden *agenda* of expectation that's currently accepted as normal…at that time. Evolution has seen those expectations altered from generation to generation. As the agenda changes, so too does the gender's presentation.

Sexual reproduction = agendered reproduction = poor copies containing hidden agendas, based on fear; skeletons in the closet; Lucifer in the underworld. Conversely, the original *mother*-nature is unconditional and loving, with nothing to fear and nothing to hide. **[Hell: see glossary]**

The word Hell literally means to cover or conceal (Oxford Concise Dictionary). And Hades means 'the unseen one' (Wikipedia). The Underworld is clearly that which is hidden. Interestingly, the word *conceal* has an etymological connection to the word *cell*. Your truth is hidden in your cells, masquerading as your family secrets…DNA.

From the first act of sperm impregnating egg, we find that the penetrating male crosses boundaries in his need to control the world around him. He must impress his mark upon others, unwittingly depriving them of their own uniqueness; their own spirit. Fear has driven him into the limiting corral of the limbic system, from where he also impresses himself upon your body tissue in the form of hormones; impulsive chemistry containing the family's hidden agenda. The Y chromosome is inherently fearful, and it's this fear that creates the generational baton of emotional dis-ease. However, with the soothing of fear by the motherly voice of mitochondria, both the Y chromosome and the limbic system will become obsolete, along with the platoon of hormonal chemical soldiers deployed by your endocrine system.

Perhaps we'll see the end of war!

The fairy tale analogy that aligns with the mitochondrial evolutionary plan that's designed to soothe the Y chromosome, is Mary Poppins.

We all know that Mary Poppins is a magical English nanny, who's blown in by the East wind. The fact that the East wind is indicative of change and progress reinforces the mitochondrial analogy. (wiki)

It's Mary Poppins' mission to soften Mr Banks by tempering his old ways, so as to liberate the children from his rigid, patriarchal rule. As the head of the family, expectation requires that he toil in the bank. He's indicative of our imaginations that are trapped in our limbic *bank* of memories.

As nanny, Mary Poppin's feminine *motherly* nature introduces magic and joy and fantastical imaginings into the family home, bringing an awakening and heartfelt transformation to Mr. Banks.

Mary Poppins admirably represents the mitochondrial Mother Goose.

What's clear to me is that the feminine, creative essence is stretching the human imagination to its innovative fullness, whilst concurrently rendering the limbic system obsolete.

Both increased innovation and limbic differences are the domain of autistics. Accordingly, the demise of the Y chromosome, along with its inherent dis-ease, must also have a direct relationship with the increase of autism. Far from being a disability, autism therefore, is the evidence of the innovative and loving presence of our *mother*-nature within the human condition. Autistics are leading the way on the spectrum of evolution, when all is seen in the right light.

The limbic village, which nestles around the brain's third ventricle, is dismantling its defensive borders, and opening the third ventricle to the free flow of creative information between the pineal and the pituitary, which was disconnected with the onset of atrophy during childhood. This is the opening of the third eye, which the gentle whisperings of your mitochondrial Mother Goose encourage as she lovingly reinstates your first and fondest stories, and regenerates the nursery of your infancy within your body tissue.

Your infancy (inner fancy) is when you were indeed connected to the eternal spring of universal knowledge, the fully open imagination, pure energy; just like geese and autistics. Mother Goose deems it so.

As the cell's powerhouse, your mitochondrial Mother Goose generates energy in story form; your chemical energy that's encoded in smell. Your mitochondria pass on your re-imagined story from generation to generation in the form of smell, making refinements to your aromatic fingerprint (DNA) in the process, whilst encouraging the optimal flow of beautiful essence through your energy meridians. The word meridian figuratively means 'the point of highest development, or fullest power'. When your energy flows freely through your meridians, you are aligned with the highest light frequencies; the eternal spring of universal knowledge; your fully open imagination, wherein all dreams come true. Your energy is radiant and infinitely attractive; magnetic; charming. Your mitochondrial Mother Goose has brought you to the Happily Ever After of your life fairy tale, where everything's coming up roses.

To continue the Lucifer analogy, we find another parallel in the god Attar, who descended and ruled the underworld according to the Canaanite religion. (Wikipedia) You may remember that the word *matter* has a close association with attar, which is the aroma of flowers, particularly roses. In its disguise as Lucifer within your fiery cells, attar has fallen from its highest floral essence of roses into the dense energetic state of your smelly hormonal chemistry, courtesy of DNA. And the resurrection of Lucifer is recorded through the increase of power in your meridians, as they channel your refined floral essence. Your body's smell becomes rosy, and you become more magnetic, in alignment with the healing patterns of your life fairy tale.

I'd like to introduce you to just one such meridian, because it so clearly reflects the autistic experience. The Pericardium Meridian (PM) is the protector of the heart. And we all know that autistics are gentle, loving souls, with a beautiful sensitivity. Mike Mandl, in his book Meridians, Maps of the Soul, writes…

> *Sensitivity per se is a beautiful quality as it represents a finely tuned discernment for people and situations, an excellent eye for details and nuances, an almost magical intuition, great empathy and lateral thinking.*

He goes on to say…

> *But walking around today's performance-oriented culture with a constantly open and soft heart is not always comfortable, either for the person concerned or for their environment, which more often than not will lack any understanding of such a hypersensitive state.*

When the physical, environmental onslaught is too much, the PM employs protective energy by dampening the sensory overload. But if the PM energy is weak and scared or defensive, the heart can be either too open or too closed. When too open, the heart can be too easily trampled on. The person feels what others feel, and experiences what others experience. Disharmony is avoided at all costs because it's a direct hit to the heart. There is overstimulation, hyperarousal, tension and overwhelm. (Mike Mandl)

> *A weakness of the pericardium meridian often goes hand in hand with timidity, reclusiveness and fearfulness, or even social phobias.*

Are you seeing the autistic characteristics?

When too closed, the heart becomes hard, rigid, inflexible, especially regarding emotional issues. These people keep others at a distance because closeness feels too dangerous. They appear emotionally stunted. (Mike Mandl) More autistic characteristics.

In both cases, the PM is being protective, but it's using two very distinct energies, which are fuelled by two very distinct background stories. One is overly fearful and protective, and the other is angry and defensive. And it's these energies that affect the body tissue and organs in their vicinity…in this case, the heart.

> *A healthy pericardium meridian is a responsible gatekeeper, a wise protector of the heart. It knows precisely what it allows through…and also what it has to keep out and when. This knowledge leads to a high degree of sovereignty and self-determination in emotional matters.*

Evolution is transforming the energy in everyone's meridians from fear and defensiveness to love and nurturing, as it re-imagines your life story into one of empowerment.

Your power is in your story; in that which you believe to be true.

Story holds power.

Story is creative.

Story is the focus of this chapter. And so far, we've looked at the maternal storyteller, your mitochondrial Mother Goose, her fairy tale equivalent, Mary Poppins, and your meridians, along with the traditional storyline of your Y chromosome. I'd now like to welcome another story into this evolutionary narrative; one that might raise a few eyebrows. In actual fact, it's a collection of stories or parables, which is inherent in its name, the Bible, literally meaning '(the) books'.

I haven't gone all religious on you, but I do love to play with words and find their hidden meanings and connections to our spiritual and physical evolutionary journey. You may also remember that one of the definitions for the word evolution was 'the unrolling of a book'...the Bible perhaps? The word Bible literally stems from the Greek for paper or scroll, with an extension to parchment, tissue, roll or helix. Is the Bible secreted within your tissue and DNA double helix? Let's get back to mitochondria so we can find out.

Effectively, mitochondria are reconnecting each and every human being to their creative higher power, as she threads her nurturing *mother*-love into their belief system, creating a more beautiful essence. **[nurture: see glossary]**

Mitochondria are powerful storytellers; powerful creators; powerhouses of love.

Higher power; creation; love; beliefs; nurturing. Are these not the characteristics of the concept known as God?

Mitochondria appear to be establishing God within the temple of your body. Many cultures around the world embrace the concept of God, the higher power and creator of all, albeit using different titles or taking on different forms. The understanding is the same. So, when I'm talking of God, I'm referring to this universal, creative, higher power. The power that I believe is within us all.

Figuratively speaking, the creative energy of mitochondria is the creative energy of God in the form of story, which makes it the *word of God* or *Bible stories,* otherwise known as the *Holy Scriptures*. So, the Holy Scriptures are your sacred beliefs; the beliefs that are sacred to you, and that you practice religiously within the tissues of your body temple. Your personal Holy Scriptures are indeed written into your DNA script within your cells. **[holy: see glossary]**

Mitochondria is your ever-present loving creator. She knows the truth, for she *is* the eternal spring; the font of knowledge; the wellspring of love; the fountain of youth and beauty. This she imparts to you through your DNA, bit by bit, via the evolutionary crawl, slowly infusing it into your unique and sacred blueprint of beliefs. Until one day your beliefs match hers, and the cycles of dis-ease come to an end, along with the evolutionary baton of DNA. You find that you've arrived at your happily ever after within the arms of your very own original *mother*-nature; your pure, youthful body temple.

The Bible story is often referred to as the greatest story ever told, however it is simply a disguise for the fairy tale called evolution.

Does not the Bible represent the progressive path of evolution; the inward shift into your original *mother*-nature? The Old Testament into the New Testament; the patriarchal into the feminine; concluding with the words 'May the grace of God be with you.'

"Hail Mary, full of grace..." Mother Mary is the epitome of the feminine, and she's also known as the Mystical Rose. Her mystery is the great revelation of evolution; that she's been within you all the time, guiding you home through her mitochondrial disguise, bestowing forgiveness upon the body tissue of humanity, as it's assumed into its highest form. Unconditional love sweeps through your energy meridians, as you're graced with the highest power through the essence of roses.

I love this little quote by C.S. Lewis, although the pronoun might need updating.

> *God is the storyteller and Providence is his own storyline.*

C.S. Lewis, cited in Becoming Mrs. Lewis by Patti Callahan (2008) (page 128)

Ultimately, your personal story teller, mitochondria, oversees the providence of your own storyline...your beliefs, which are indeed the foresight that creates your reality. Your beliefs are the mothers of your imagination; the wings upon which you can fly miraculously beyond all limitation into the eternal spring of possibility.

Before leaving the Bible, I feel that it's important to address its main character. Jesus is known to be the light of the world and *the word of God*. As stated earlier, it's the word of God that expresses through your mitochondria in story form, therefore Jesus figuratively lives in your cells, bestowing the miracle of life upon your physical body. **[Jesus: see glossary]**

CHRIST JESUS = X JESUS = EXEGESIS.

Your exegesis is your creative expression or the out-pressing of your uniqueness, your spirit, your little light; your individual gene-expression; your DNA. Having been crucified by a rigid external world, your unique expression is now being resurrected from its dark tomb of fear. *You* are the light of the world.

> *Let the universe show off its creative flair for the unexpected and inspirational through you.*

Alana Fairchild (2017), Love Your Inner Goddess oracle cards

In pushing the boundaries of the imagination, your mitochondria are engineering the inward shift of evolution that's leading you all back to your original and creative genius...your higher power...your loving, motherly voice of God. Autistics have simply spread their wings sooner and travelled higher, giving the illusion of disability to those who cannot yet see the beauty in this great plan of evolution.

Interestingly, mitochondria have been implicated in several human disorders and conditions, including autism.

Given that the limbic brain is where mammalian evolution is recorded and that your mitochondrial whisperings are responsible for these changes, it stands to reason that your limbic system is collaborating with your mitochondria. Throughout evolution your limbic system has been providing the appropriate measure of defence in alignment with your family stories (DNA). As your mitochondrial Mother Goose progressively increases the presence of her loving, feminine voice within those stories, your limbic defences diminish. You don't need a defence system in a loving story. You don't need a defence system in a loving world.

Therefore, the apparent under-functioning limbic system of autistics, once again, provides evidence for a more highly evolved mitochondrial Mother Goose. Their defences are down because their original *mother*-nature is more present.

The inward shift of evolution is made possible through the creative power of story. Your inner genius is emerging through story. Start listening to the stories of autistics; you might be surprised.

Instead of seeing autistic mitochondrial difference as dis-ease and disability, just because it doesn't align with expectation and the norm, ask yourself, 'what if the autistic mitochondrial difference is the result of a more highly-evolved story coursing through their brains?'

If you don't, you're suppressing their intuitive evolutionary narrative; their genius; their guiding spirit; their generative power; their power of creation. Can you imagine how frustrating that is?

Yes, story is power. And the autistic story is higher power.

Not through the force of muscle, but through the magnetic influence of high-frequency energy; love and light; ingenuity and imagination.

Muscle is the masculine, or wounded feminine. Muscle is power-*over*, not empowerment. Muscle is reptilian, patriarchal, the Y chromosome. It's the WHY? chromosome that's so afraid its ignorance will be discovered, that it builds muscular walls of defence, in the name of masculinity. Muscle harbours fear.

The Oxford Dictionary of Etymology defines muscle as a contractile fibrous bundle producing movement in an animal body. Figuratively, muscle is an emotive bundle of tradition that's held together firmly by proteins. Tradition flexes its muscles within the defensive walls of your protein structure. .

[protein 1 & 2, polypeptide: see glossary]

> *Excluding water and fat, the human body is made up almost entirely of protein. Protein is the main component of muscles, bones, organs, skin and nails.*
>
> *Excluding water, muscles are composed of about 80% protein*

Proteins: building blocks of the body. Otsuka Pharmaceutical Co., Ltd.

A protein is a polypeptide chain. By waving my word-magic wand over the word polypeptide, we find that poly means many, pep means life and tide relates to time. Now by putting it all together, we can see that polypeptides are *many life times*. Your protein polypeptides embody many life times of habit. Many life times of family tradition are held within your protein polypeptides. Your ancestors are flexing their muscles beyond the grave within the fibre of your being.

It's clear that your protein-filled body matter is a habit, with its instruction stemming from the family traditions of your DNA, as recorded in your limbic system. Therefore, as your mitochondrial Mother Goose infuses her increasing feminine presence within your DNA, the loving habits of your original *mother*-nature become more apparent in your protein structure. Less inherited dis-ease begets greater ease and increased health and well-being; a slowing down of the aging process; a nurturing of youth within your body tissue. Leading, of course, to the increased longevity that's evident throughout evolution; and ultimately to immortality. **[magic: see glossary]**

Your limbic systems DNA contract is softening, and so too are the contractile fibres of your muscles. As the wounded feminine is soothed, she loses her masculine edge, along with her defensive muscular presence. **[masculine, muscle & pain: see glossary]**

Autistics are renowned for having low muscle tone. Studies on autistics have shown that their gross motor function is significantly related to their executive functions. (Li, wang, Xin, Gu) In other words, their muscle/protein responses, or muscle-memory habits, relate to their limbic system, where executive function is based. For autistics, their more highly evolved limbic system exhibits a more loving, feminine essence, thus reducing their muscle response; reducing their masculine reactivity; reducing emotions.

And so, the low muscle tone of autistics provides even further evidence that they are indeed forerunners of evolution, as their muscles relax to the soothing strains of a story that feels loving and true.

Without the muscular tug-of-war, the skeleton also relaxes, allowing the mitochondrial story of truth to build within its fibres.

$$\text{mitochondria} = \text{thread} + \text{cartilage}$$

Mitochondria and your skeleton are intimately aligned, as evidenced within the word itself. And the limbic system is the mediator or middle-man, through which mitochondrial refinements are delivered to your skeleton. Throughout evolution, your mitochondria are threading the story of truth, otherwise known as the fullness of integrity, into the fibres of your skeleton, as she refines your limbic story. This progressive refinement is emulated within your meridian system, increasing your energy flow and magnifying your magnetic power. And so, your skeleton, as the emblem of death, is slowly but surely resurrecting and returning to its original infant form and function as the fountain of youth.

Your infant skeleton is cartilaginous, before the onset of bone. The light of truth radiates as youth through pure infant fibres. *Mother*-nature reigns joyfully as your mitochondria ensure the integrity of your cartilaginous skeleton. Her gentle voice threads your first and fondest memory into the weave of your tissue, lulled by the loving energy that flows through your meridians. Your radiant little light holds your higher power; the story of creation; your magnetic genius.

This is your beginning…and this is also your ending; the goal of your evolutionary journey; to return to your true story and shine your light through your radiant skeleton. The Alpha and the Omega.

It is your light that lights the world.

Jalaluddin Rumi

To help with this explanation, I'm going to return to my Phosphorus/Lucifer analogy from page 154, particularly in relation to their light-bringing characteristic. You may remember that Lucifer, as the fallen angel, is indicative of your infant body having fallen from its light-filled, angelic state into the fiery, emotional cells of adulthood. As Lucifer's chemical equivalent, phosphorus is in bondage to oxygen, and is unable to shine its light.

So how does the skeleton fit into this analogy?

Interestingly, about 90% of phosphorus is stored in bone. (Michigami. Ozono) That would mean that Lucifer resides predominantly in your skeleton, and that the fiery inflammatory state of bone epitomizes Hell. Throughout childhood, your fearful, patriarchal limbic story has caused your heavenly infant skeleton to deteriorate into its bony state of Hell. Fear is holding phosphorus in bondage within your skeleton. Fear petrifies the flow of energy through your meridians. Fear creates the smelly essence of hormones in your cells. Fear creates mortality. Fear is the Grim Reaper. And isn't fear the predominant emotion in images of Hell, which often depict the Grim Reaper and ailing skeletons within a fiery pit?

Your skeleton is in need of love. All your connective tissue is in need of love.

And that's exactly what evolution is achieving, as the mitochondrial Mother Goose slowly infuses more love into your every thought, word and deed. It's *your* loving behaviours that are encouraging the oxygen bonds to loosen their hold on phosphorus, so it can release its angelic light back into your skeleton. It's your love that's causing Lucifer to rise out of Hell, as you nurture your bony skeleton back to its original, light-filled cartilaginous form. It's your love that's fostering the return of attar, the essence of roses, within the fibres of your skeleton. It's your love that flows through your meridians, empowering your body. Love is the power that's been magnifying its presence throughout evolution, both within your skeleton and all your connective tissue.

Autistics' connective tissue narrates their more highly evolved, softened mother story; the gracious feminine. They're not enamoured by the rigid and limiting system of neuro-typical expectation with all its strings and controls; the masculine. Their skeletons are no longer puppets dancing to the push and pull of bossy over-sized muscles; their under-developed musculature a fading footprint of an antiquated regimen.

Your mitochondrial Mother Goose relays the story of evolution through your skeleton. Your light of truth is no longer a mystery to your carbon fibres when the fairy tale of evolution reaches its conclusion.

Evolution is written in your bones.

Your story is written in your bones.

It's time to release your skeletons from their dark closet of secrets, so your unique creative genius can shine.

This is what autistics are asking of you: please let us shine *our* light. Let us narrate *our* story. It's written within the fibres of our being, for all to see. And it's a wonderfully imaginative story of love and light and creative ingenuity.

Don't change our story to match yours. Open your imaginations to see the true fairy tale of evolution that's unfolding within your mother temple.

We are seeking in the cathedral of bones for the poetry of oneness

Gabrielle Roth

Chapter Ten: Denouement

...the solution of a mystery, the winding up of a plot, the outcome of a course of conduct.

f Fr. **denouement** "an untying" (of plot) from ***denouer*** "untie"

f ***des-*** "un-out" + ***nouer*** "to tie, knot"

f L ***nodus*** "a knot", f PIE root ***ned-*** "to bind, tie".

Online Dictionary of Etymology

The Dictionary of Oxford Languages elaborates upon the meaning of denouement thusly,

...the final part of a play, film or narrative in which the strands of the plot are drawn together and matters are explained or resolved.

And so it is that your mitochondrial Mother Goose reaches the denouement of the evolutionary fairy tale, in which the strands of the human body are literally drawn together as its cells become clear of their long-held prisoners of mystery...your chemical personas. When your masks are removed, the truth becomes crystal clear, and your *mother*-nature steps into the spotlight for her final curtain call, draped in the light-filled, rosy fabric of immortality. **[chemistry: see glossary]**

As the narrative reaches its happy conclusion, the limbic system bids farewell. The *ties* between memory and emotion are finally broken. TIME stops! And your aging tissue remembers her true, youthful nature. No more ties...no more limbic.

It's the limbic system that's being *untied* throughout the process of evolution. The disappearance of the limbic system marks the denouement of evolution...happy news for autistics, as their under-functioning limbic finally makes sense. Once the fear-filled *ties that bind us* (DNA) are overcome, mammalian evolution will have reached its conclusion, as the course of family conduct embraces its final outcome...full rehabilitation of family habits; meaning there's no more family dis-ease from the past to be handed down to future generations.

No more past or future. Just a beautiful *present*...a gift of love. **[meditation: see glossary]**
When unwrapped, this is the gift that is always present, awaiting your choice. Being in the moment provides you with the opportunity to choose *loving* thoughts, words and actions. The present moment is a meditation of love that fills your tissues with the feel-good vibes of your light-hearted spirit. Your *mother*-nature glows as she nurtures her wondrous, magical and imaginative spirit of youth, renewed moment by moment in her eternal hug of love.

As each individual opens their heart to receive this gift of love, and stands in meditation within their own *mother*-nature, they'll awaken to the understanding of what it means to ride the high vibes. And with this revelation, they'll stand side by side with autistics. This denouement of evolution will bring liberation for autistics, as they'll finally be understood.

Now, let's step inside the body once again, so we can discover where the happily ever after of the evolutionary fairy tale takes place. On this final leg of our journey, we're going to pull everything together as we traverse the length of the spine and into the brain; from base to crown chakras; from the coccyx to the corona radiata. Your *Mother*-nature will reveal her true physical design. Having transported many cycles of *reproductions* through the evolutionary cleansing, her queenly *original* will finally emerge, triumphant, wearing her radiant crown.

The butterfly of transformation will lead you through the secret garden of delights that's been patiently awaiting your return at the tip of your spine...your ovum. She is the *mother*-of-all-cells; your spring of life; your font of knowledge; your fountain of youth; the source of creation; the playground of evolution. Within her loving heart, your radiant little light, your spirit of joy, ignited her imagination in the beautiful Once Upon a Time of your unique life fairy tale. And in this cocoon of love she's swaddled your light and spirit, watching over her precious treasure, while you detoured into the trials and tribulations of this wayward world. You've always been slumbering in your mother's arms, while you lived this dream called reality. In spiritual circles, this sleeping energy is known as the kundalini; within religions, she's the Holy Spirit.

As the kundalini, she's represented by a coiled serpent that's indicative of your latent spiritual energy; your infinite potential; and joy. She's simply the creative power of your *mother*-nature; your loving feminine; your magnetic attraction; your imagination. Her creative essence lay helplessly dormant within the serpentine coils of your DNA. **[kundalini, latent and desire: see glossary]**

Throughout evolution, your kundalini has been sitting in your ovum, at the tip of your spine, slowly but surely rising out of her creative slumber through the ongoing expression of innovation; the unfurling of genius; the gentle whisperings of your mitochondrial Mother Goose. She's struggled through the ignorance of a fearful, doubting world. But she's persevered nonetheless. Kundalini has been rising, in all her glory, through the awakened spirits of pioneers, innovators, pattern-seekers and imagineers, who've graced this planet down through the ages. And it's this persistent rising that's now finding her voice through the creative genius of autistics.

The *kundalini rising* is an oft heard spiritual term that's been aligned with the concept of an ascension into higher realms. Clearly, however, it also has a physical/scientific counterpart in the evolution of the human body's energy system.

Throughout evolution, the kundalini has had a direct relationship with body chemistry. As your kundalini rises, increasing the presence of your true creative spirit, your chemistry dilutes in a slow fade that reveals the rosy face of your pure organic carbon nature. Carbon is kundalini risen. **[chemistry: see glossary]**

Kundalini is indeed the *mystery* that's been veiled by the instinctive chemical prisoners within your cells. She's been hiding behind the masks of smell that your aromatic fingerprint (DNA) has dictated. When kundalini rises fully, your cells are cleansed and there's nowhere to hide, for you stand in your truth. When you take back your power, everything comes up roses. Kundalini is empowering...gentle and motherly.

With a slight shift in perception, it can be seen that your ovum is the power point of your physical body, as she carries your kundalini; she's your source of creative energy; your storybook of life.

Interestingly, Freemasons call this point your *circumpunt*. Symbolised by a dot within a circle, it perfectly represents the word itself, circum (circle) + punct (point). However, this also translates into *circuit point*, the point of power or power point; the infinite source of power or first cause of creation. And as the *cycle point* it becomes the point through which the *cyclical* generations narrate the story of evolution in round after round of life times, wherein your mother-of-all cells receives her precious cargo.

In Freemasonry, it sits at the base of the spine, where they also place the Holy Scriptures (Bible), and it denotes having attained the level of Master Mason, or the 33rd degree. (the word *degree* means *down step*) So the circumpunt's physical counterpart is the ovum, which sits at the 33rd vertebra, or the *33rd step down* from the top of the spinal-staircase, and which contains the Holy Scriptures of your DNA that are being refined through evolution. **[circumpunct, master mason, freemason: see glossary]**

When you've overcome fear and petrification, you'll have *mastered* your physical body and will no longer be working with the stony, petrified tissue that accompanies the aging process. You'll have mastered the masonry of fear. And your body tissue will be as youthful and soft as it was in your infancy.

The circumpunct is also known to be the alchemical symbol for gold. This leads us to another beautiful analogy, for when you've refined your Holy Scriptures/DNA, your kundalini will rise through your spine, uniting your seven (rainbow) chakras into a beam of white light. No more rainbow.

Your radiant, motherly ovum has transformed into the pot of *gold* at the *end* of the rainbow (when the rainbow has gone).

And further to that, the circumpunct is the symbol for the centre of infinity, which is otherwise known as the emanation or first cause. You'll have reconnected with your first and fondest infant/infinite nature; the universal field of unlimited possibilities; the fully open imagination; the source of creation; your heart of infinite joy.

Like a fountain of light, your kundalini rises through your spine to effervesce her inspired imaginings joyfully through your third eye. Your sparkling spirit celebrates her liberation; her full revelation; her creative flair.

This full revelation of your creative spirit is also called the Holy Spirit, flying free of all limitation, having cleansed the past from your cells, through her long-running epic of forgiveness. She's your full inspiration, the essence of truth and integrity. The Holy Spirit is said to come down upon you, simply because she soars on high, amongst the highest frequencies. Your ovum is the baptismal font of evolution. **[Holy Spirit: see glossary]**

Baptism, or Christening, is also a Naming ceremony. Names are simply words, and all words carry vibrations, whether they're thought, spoken or written. The evolutionary baptism is, therefore, a purification of words; the words that narrate your story; the words you think, say and write; the words you believe. **[pineal 1: see glossary] [it's necessary for all to read this definition]**

It might seem strange to reference a bible character in relation to human evolution, but it seems obvious that John the Baptist would feature in this long-running epic of *forgiveness*. Once again, I'm not going all religious on you. I just wish to impart this fascinating analogy, and perhaps further illustrate that the Bible is saying more than you think.

Figuratively, *John the Baptist*, being *the gene that adapts*, relates to the epigenetic influences that evolve the physical body by cleansing the cellular information, making way for higher frequencies and increased light. Can you see your mitochondrial Mother Goose in this literary refinement?

Epigenetics *cleanses and purifies* the word...and the body.

John the Baptist *baptised* Jesus, who was The Word made flesh.

Mitochondria *refine and purify* your story/words...and body tissue/flesh.

Mitochondria cleanses your DNA and forgives the past. Mitochondria immerses the masculine in feminine energies of love and light. Mitochondria enlightens, or brings to light, your unique creative expression, and the purity of your eternal form. Both mitochondria and epigenetics are personified by John the Baptist. **[water: see glossary]**

Now to the final interesting morsel that figuratively connects mitochondria to John the Baptist, and both to human evolution. One of Leonardo da Vinci's paintings of John the Baptist, shows him holding a staff in one hand and pointing upwards with the other. In alignment with his epigenetic duties, the upward point indicates the ascension into higher frequencies, otherwise known as evolution. And the staff represents mitochondria, which has bacterial origins; the word bacteria meaning *little staff*. I don't think it's a stretch to say that Leonardo was clearly ahead of his time in his understanding of the workings of the human body...and of evolution. **[bacteria: see glossary]**

John the Baptist was the forerunner of Jesus Christ. Mitochondria are the epigenetic forerunners of the body's ascension into higher frequencies of light through cleansing and evolving DNA. Autistics are the forerunners of evolution, shining their bright lights for others to follow.

As I've explained in previous chapters, the third eye of autistics is open and their pineal is leading the way into the unexplored territory of evolution through innovation and imagination. Their enlightened mitochondrial story is evident in their beautiful, loving and forgiving *mother*-nature. The fullness of their unique spirit has come down upon their cells; their Holy Spirit shines; their kundalini has risen.

The most obvious clue that autistics are the forerunners of evolution is actually embodied in their *label*. Autism is derived from the word *auto*, meaning *self*. Kundalini rising denotes *self-realisation*, which is the fulfilling of one's own potential. Autistics are fully filled with the light of their unique spirit, which they nurture as a mother does an infant, even to the point of withdrawing from this dense world, if that's what it takes to stay true to themselves. **[autism: see glossary]**

When you're true to yourself, and shine your unique light, the flow of higher frequencies fills your body tissue, which reinstates a channel through the centre of your spine. Called the lumen, it's the adult remainder of the embryonic neural tube that closes with age. It's no coincidence that the word *lumen* means *light*, for it truly is a channel of light, when fully open. During infancy, it transports the light from your ovum, or eternal spring of knowledge, through to the third ventricle in your brain, signalling that your third eye is open. This channel of light is also called the hyaline canal, and is synonymous with the spiritual sushumna. **[hyaline canal: see glossary]**

> *In the fetus,…a canal, continuous with the general ventricular cavity of the brain, extends throughout the entire length of the spinal cord, formed by a previously open groove.*
>
> *In the adult, this canal can only be seen at the upper part of the cord,…extending for about half an inch down the centre of the cord, where it terminates in a cul de sac, the remnant of the canal being just visible in a section of the cord…its cavity having been obliterated.*
>
> *Gray's Anatomy page 191*

With the severing of the hyaline canal, before birth, your creative power source (ovum) becomes disconnected from your brain's ventricular system, and the light of your higher spiritual vision dims. Your third eye closes. However, your ever-inventive *mother*-nature, intent on liberating her sleeping charge, immediately creates an alternative, although inferior, postal system through which to channel messages to her connective tissue. Necessity is the *mother* of invention after all…her needs are evolutionary.

The demise of your unified light *channel* has necessitated the formation of your dualistic human *race*, which relays messages through the central nervous and endocrine systems of your spinal cord and brain. Within this new system of fear, the left and right sides of your body and brain operate separately, with cellular information now expressing reality through your bifocal vision that deteriorates with age. Your dualistic five human senses have replaced the higher sixth sense of light as the means of energetic communication. And your limbic system sets up its protective enclosure around your third ventricle, to assist your *mother*-nature in her evolutionary pursuit back to her origins.

Nestled around your central nervous system (and your pineal gland) are three layers of membranes called the meninges. As their names suggest, the Pia Mater, Dura Mater and Arachnoid Mater act as protective *mothers* to your brain and spinal cord throughout your evolutionary journey. They are assisting in the refinement process that's transpiring in your neuro-endocrine systems, and will continue their nurturing role until you've fully evolved. The three meningeal mothers are integral to the kundalini rising within your body, for it's through them that you reconnect to your ovum, which is your power source or point of infinity.

As the mother energy, the kundalini is represented by 3 and a half coils. In spiritual terms this refers to the 3 gunas (mothers) plus the transcendence to infinity. Physically, it's the three meninges and their connection to the ovum.

> *...the half representing infinity, no ending and movement towards greater and greater expansion, reminding us that this energy represents our own infinite nature.*

Kundalini; the serpent power (2023), The Sattva Collection

At the base of your spine, the fibres of your meningeal mothers draw the evolutionary narration together into its final thread, the filum terminale, through which the aforementioned central canal will reconnect with the coccyx and ovum, reopening your loving mother-infant dialogue (matter-infinite).

The Online Dictionary of Etymology tells us that the word *filum* derives from the Latin word for *thread* (just like the *mito* in mitochondria). And so, the filum terminale brings us to the termination of the thread, or the end of the yarn/story, of evolution; the fairy tale happily ever after.

> *The filum terminale is the nonfunctional continuation of the end of the spinal cord. It usually consists of fibrous tissue without functional nervous tissue.*

N Gupta et al. (2017), Disorders of neural tube development Filum Terminale - an overview

The filum terminale has no functional nervous tissue; no smelly fear-filled threads. Just a loving conclusion.

You may also hear the word phylum, meaning a race or tribe who are genetically related.

Accordingly, the filum (phylum) terminale marks the end of the race as we cross the finish line of evolution, as well as the end of the *human* race, as dictated by the tribal nature of DNA.

I've previously explained that the human *race* is the communication *channel* that sits between the base and crown chakras of every individual, in the form of the central nervous system and brain. It's in this central channel that the transformation from your mortal existence into immortality is taking place, with the evolutionary transition from the illusion of fear into a new reality of love. **[evolution 2 & illusion: see glossary]**

With the ever-increasing presence of your *mother*-nature, it becomes clear that she is your true and original essence, and that fear is indeed an illusion that you've simply bowed down to blindly. The disconnection from your loving mother-cell has literally created the *ill vision* of your human dualistic sight, with the closure of your third eye. But, as the glossary definition states, evolution is bringing you out of this illusion of fear, and back into a higher vision of love.

Your transformative *mother*-nature is present in every tissue of your body; in every cell membrane. She has a front row seat in your physical and spiritual evolution. **[membrane: see glossary]**

Cellular biologist, Bruce Lipton, tells us in his book The Biology of Belief, that cell membranes are the cell's equivalent of a brain.

> *The membrane's function of interacting "intelligently" with the environment to produce behaviour, makes it the true brain of the cell.*
>
> *Your cell membranes are information processing systems with protein molecules embedded in their oily inner layer, relaying information into your cells that determines their function, and subsequently your behaviour...and perception.*

> *Protein complexes are the fundamental units of cellular intelligence...referred to as units of perception.*
>
> Bruce Lipton (2005), The Biology of Belief: unleashing the power of consciousness, matter & miracles (page 87)

But proteins are simply habits; stored memories; old beliefs; smelly triggers. These antiquated stories have literally embedded themselves in your cell membranes perpetuating the illusion of fear, like ancestral family portraits lining the walls of a castle. Your patriarchal traditions have taken control of your creative feminine brain as her essence falls foul to his instinctual dictates. Her creative nature diminishes as her fabric is pierced with little bundles of limitation. And this is what science calls intelligence!
[intelligence 1 & 2: see glossary]

Intelligence is a habit, as your smelly proteins drive your cellular functions and responsive behaviours; your limited and predictable perception.

Science sees this intelligence-mechanism as normal. But what if the invading proteins disappear so that the environmental information can no longer penetrate the cell and trigger emotional reactivity and deterioration? Now that sounds intelligent to me!

When seen through an innovative lens, it becomes clear that this is exactly what evolution, in her wisdom, is doing. Through the progressive refinement of habits, evolution is raising the collective intelligence levels of humanity, and reinstating the *creative* nature of your cellular membranes, thus bathing them in increased light and youth and the pure essence of roses.
[evolution 3: see glossary]

From the glossary definition, it can be seen that evolution is the long-running process of expressing a progressively more purified smell, through the refinement of your aromatic fingerprint, DNA, which determines your proteins. It's the smelly habits of your proteins, embedded in the oily-layer of your cell membranes, that are being evolved, refined, purified...and laid to rest. Every cell membrane is remembering her youth.

Start listening to your mitochondrial Mother Goose as she gently refines your story and invites you to change the way you interact with the environment. She's reinstating your rosy *mother-nature* within every cell membrane of your body. The antiquated patriarchal programming that's packaged into protein bundles within your cell membranes, is weakening its hold. When you stop blindly following the patterns of automatic reactivity, you are no longer a prisoner to the environment or the past. That's when you'll be released from your cells.

Your original cell, or ovum, is patiently awaiting this happy conclusion. For she is the queen of your bodily kingdom. When her own plasma membrane was pierced by the sperm, it set in motion the pattern of protein invasion in all her cells. So, it stands to reason that her queenly plasma membrane, therefore, is your original creative feminine brain, holding the highest-frequency aromatic bouquet. Over her plasma membrane, your ovum wears a coat-of-light called the zona pellucida, which is then surrounded by her corona radiata. In her true design, your loving queen wears a cloak of light and a radiant crown of roses.

[corona radiata: see glossary]

The ovum's corona radiata cares for her after she's been released from the ovary, but is degraded by the digestive enzymes of the sperm during the fertilisation process.

I find it interesting that both the ovum and the brain have a corona radiata.

The brain's corona radiata is a crown of white matter that sits inside the dome of the skull. It carries most of the neural traffic to and from the cerebral cortex. The fact that the nervous system is a system of fear, responding to external stimuli, tells us that the brain's corona radiata is trafficking fear. Clearly, in its human guise, it's operating as part of the inferior postal system that was established with the closing of your central canal and the disconnection from your ovum.

To use a biblical analogy, your crown of roses has transformed into a crown of thorns. **[spine: see glossary]**

Your original queenly ovum wears her garland of roses as a radiant crown, relaying her high frequency essence through your open central canal, beaming through the rose-window of your third eye, and creating a saintly halo of light within your brain's corona radiata. But, upon its closure, the inferior postal system relays instead your *nervous* information through your spine, creating a crown of thorns of your brain's corona radiata.

Evolution is transforming your crown of thorns back into its original garland of roses, through the process of *overcoming* difficulties and perplexities. As you *come up over* fear into the higher frequencies of love, you transform your body into a cathedral of light, bearing a beautiful rose window of pure motherly essence that radiates from your pineal.

Concurrently, your bony frame releases its weight-bearing duties to cradle the beautiful light and rosy essence of your unique spirit, which was once the radiance of your pregnant mother.

Harking back to Chapter Eight, we find that one of the definitions for the word *skeleton* is *mummy*. The embryonic, infant skeleton carries the feminine essence of your original *mother-*nature, in her unwounded form. In its true design, therefore, the fibres of your skeleton are conduits of light; enlightened carbon. Choirs of angels. (choirs = C wires = carbon wires) Within the ovum, your skeleton is entrusted with the high-frequency blueprint that's pregnant with the full integrity of your body.

The seed has been sewn in paradise; in the ovum's garden of delight...holding in its precious heart the complete instructions for your whole (holy) spirit and enlightened, youthful body.

Until the slippery reptilian, who's been slithering through the family tree throughout evolution, reveals the fruit from the tree of good and evil; the ancestral tree of feminine and masculine; your DNA, bearing the mask of your persona. And so, your human, patriarchal blueprint establishes its dualistic form instead, sending your skeleton into its dark, evolutionary slumber through the closure of your spinal canal. **[devil: see glossary]**

In alignment with the biblical Eve, your original mother (ovum) consumes this fruit of duality in round after round of childbirth. Each time being banished from the original garden of delights. Each time dimming your skeleton as she fades into mortality, wearing her thorny crown of death.

But each time, the ovum is cleansed just a little more by the whisperings of your loving mitochondrial Mother Goose. Until one day your skeleton once again overflows with delight. **[delight: see glossary]**

You've returned home, to be wrapped in the beautiful aromas of your own rosy garden, having fully opened your heart (the canal that rises through the *centre* of your being), and therefore your body, to the delights of pure joy and love; the wonders of the universe springing to life through your pineal, in a burst of dreams and magic and imagination. Your brain's corona radiata lights up, a heavenly halo, befitting your new queenly reality, your skeleton a flowing gown of light.

Kundalini has indeed risen and there's nowhere to hide, for you stand in your truth.

The fairy tale equivalent being Alice in Wonderland. Its author, Lewis Carroll, who'd studied Greek philosophy at Christ Church (Wikipedia), knew exactly what he was doing when he created this favourite of fairy tales, to be passed down through the ages from mother to child within many a nursery.

For Alice means truth or disclosure, stemming from the Greek Aletheia. It is also translated as unclosedness or unconcealment, with its literal meaning being 'the state of not being hidden'. It is the opposite of oblivion, forgetfulness and concealment, and therefore the antithesis of Limbo and Limbic. **[Alice: see glossary]**

Alice represents the kundalini, your truthful essence, your open and free imagination, your loving *mother*-nature, expressing fully within the rosy garden of your physical wonderland.

Alice's adventures in Wonderland are indicative of truth finding its way through evolution; Mother Goose delivering integrity to your *mother*-nature (matter).

The journey from your ovum's secret garden at the tip of your spine, to your radiant pineal and rosy crown that sits within your skull, takes but an instant. Concurrently, it's a momentous revelation that's endured the course of evolution. Your *mother*-nature is victorious.

It could be said that your *mother*-nature has been assumed into heaven, analogous to the biblical Mother Mary, or Mystical Rose; she's a mystery no more. Or that your Christ light has resurrected into an eternal Easter; an infinite Spring of light resting at the base of your spine; the original Easter egg (oestre egg).

Perhaps a sparkling Christmas tree (pineal) adorns your third eye, draped in magical dreams, and guided by the rosy nose of Rudolph (rude = red; olph = olive/oil/essence = rose essence). The light of Christ has been born into your twinkling skeleton; gifts, wrapped in ribbons of love, sit as the foundation of an eternal Christmas. Joy to the world. **[Christmas tree & Rudolph: see glossary]**

Fairy tale fantasy might be sitting in your cells as Cinderella arises from the ashes to ashes, dust to dust cycles of servitude. For Cinders is *carbon*, your body's pure organic building block. And Ella is *she*, your creative feminine. Youth and beauty meet in the castle of your dreams, reigning lovingly over your kingdom of light.

Evolution appears to be captured in every story...fairy tale or biblical. But then, evolution's own definition told us that it was the unrolling of a book. And isn't it often said that we all have a book in us.

As I arrive at the end of *this* book, my wish is that I've piqued your curiosity, and inspired you to open your own animated book of wonders and magic and uniqueness. Nestle into the universal library of endless imaginings, dreams, beauty and love, which beckons from within. Autistics are waiting there for you, although you may not recognise them without their labels; because, after all, they're just like you.

Addendum: The Facts Unmasked

An Autistic Reality

> **Fact** (n.) *f.* L *factum* "anything done, occurrence, state, condition, circumstance" [the modern, empirical sense of] "thing known to be true, a real state of things, what has really occurred or is actually the case."

So here are the facts...what is actually the case for autistics.

In this addendum, I offer simple explanations for many of the characteristic behaviours attributed to autistics. There are many more, but this is a good start.

1. MASKING

> **Mask** (n) *f.* Fr *masque* "a false face; covering to hide or guard the face; spectre; disguise."

Everyone pops on a disguise when they step out to meet the world, whether they're aware of it or not. This includes neurotypicals who don their well-crafted personas in order to participate in the social dance of life. So, what we call reality is effectively a masquerade.

Your persona is your false self...a concept first developed by English psychoanalyst and child psychologist, Donald

Winnicott.

> *It's formed as an adaptive, learned, defense to the true self...Winnicot realised that a false self is necessary in some situations to fit into society.*
>
> Darlene Lancer JD, LMFT, *The Co-dependent False Self*

When autistics *mask*, they're simply joining the *masquerade* as they mimic neurotypicals, in the name of social acceptance.

Everyone is masking.

What sort of society are we living in when it's *necessary* to be untrue to ourselves; when we're asked to hide our true identity; when the only way we can belong is to be false? What does this achieve? Heartache, anxiety, depression, meltdowns, isolation; a condition that's come to be known as autism.

But *is* there really such a thing as autism? Are we not just human beings who simply wish to be true to ourselves; to express our true essence; to be kind and gentle; to shine our own lights, in our own way...without fear of retribution? Are we not simply human beings who no longer want to wear masks in a false world that places societal dictates ahead of truth?

Why *is* it necessary to *fit in*? Who benefits? Nobody. Society is not better off when everyone's squeezing themselves into the same mold, because this fosters discomfort and dis-ease, the price paid for pretending to be what you're not.

Masking is *not* an autistic construct. It's a protective, neurotypical habit that's thankfully being rehabilitated through the process of evolution. The fact that neurotypicals display a more highly active limbic brain indicates that they're still presenting their false face to the world and masking their truth.

They're still evolving. Autistics, as forerunners, are showing the world that it's possible to remove the masks and allow the truth to shine. Can you imagine what a beautiful world that would create?

2. BEAUTY = BE YOU -TY

BE YOU and your true light shines. This is what autistics want, for both themselves and the world.

Remove your mask and reveal your beauty. Evolution will thank you.

3. DEPRESSION

Autistics are tuned into higher frequencies. This is neither good nor bad. It's simply the way evolution has dealt the cards. Accordingly, autistics are the recipients of all the information that rides these high-frequency light waves, including their unique perspectives, pattern recognition, creative problem-solving abilities, remarkable powers of concentration and gentle nature. They see the beauty in everything around them, as Mother Nature's form and function unfolds with ease before their eyes. Their hearts are filled with love, which unites all on the higher frequencies.

The fact is, this is their norm. Autistics feel at home here. Feelings of joy and ease accompany their high-frequency reality, for this is their comfort zone; their safe space.

To tune into this high-frequency world requires what I like to refer to as a beautiful sensitivity, which all autistics possess. Their finer sensitivities match the higher offerings. They're not *too* sensitive; there's no such thing, for you can only be as sensitive as it takes to be you, otherwise you'd be someone else.

Effectively, the depression that's so prevalent amongst

autistics is caused when they're unable to be themselves; to carry, unguarded, their beautiful, loving natures into their daily lives, because of the disapproving world around them; when their natural proclivities are not respected; when the people around them are not loving or lack integrity.

When autistics have to wear a mask in order to fit in, they feel sad. They know how beautiful the world could be, and it breaks their hearts when this is not reflected back to them. When the love and beauty of Mother Nature is not evident, autistics' spirits are depressed. **[depressed: see glossary]**

4. Eye-Contact

The eyes are the windows to the soul. (proverb)

Eyes don't lie. They convey the truth of how a person is feeling, even if their social mask is firmly in place. Your eyes betray your emotions and bare your soul to the world. This causes problems for beautifully sensitive autistics, with their heightened receptivity, for they *feel* the heavy load of emotions pouring from the eyes of others. The intimacy of this emotional connection makes them uncomfortable. And due to their gentle nature, it saddens autistics to know that the person in front of them is suffering.

However, this isn't the only reason for discomfort, and possible confusion. Most often, the neurotypical with whom the autistic is communicating will be wearing a mask that fits the particular social situation. Therefore, the autistic will be receiving two distinct energetic signals, one indicating the feelings of the neurotypicals' true self, and the other indicating the energies of their false social face.

This throws the autistic into a quandary. Processing two

different, often contradictory signals, and formulating an appropriate response, takes time. Awkward silences can ensue, with the autistic being seen as socially inept. Breaking eye contact at this point helps the autistic by reducing the incoming information that's causing the traffic jam, so they can focus on the required response.

I feel certain that neurotypicals also receive mixed messages when communicating with other neurotypicals, but their social programming is such that they defer to the external mask and respond accordingly. And perhaps at a later date they may recall having a funny feeling at the time, or a niggling suspicion that they put on the backburner.

Neurotypicals could also feel threatened by the ease with which autistics can read their emotional truth. However, there's no reason for concern, for the high frequency abode of autistics is very loving. They don't wish to judge or hurt you in any way...that sort of behaviour is simply not on their wavelength. All they ever want is for others to feel good, to connect through love, to offer kindness. If you look into the eyes of autistics, you'll feel their gentle nature and realise that you're safe.

Autistics don't feel threatened by other autistics because they're on the same wavelength. They recognise the loving soul with whom they're connecting. There's greater understanding, fostering increased ease. **[understanding: see glossary]**

5. BLUNTNESS

Truth, honesty and integrity are paramount for autistics. And all too often, they respond honestly, only to find that they've said the wrong thing...according to neuro-typical measures. This is why autistics are seen to be blunt. However, it's simply the result of living in a world that uses social masks to disguise the truth. Where neurotypicals *know* the rules of what to say and when, autistics *feel* the energies of the moment and respond accordingly.

Neurotypicals operate from their programmed thoughts and emotions, whereas autistics operate in the immediate, from their feelings.

Bluntness could be translated as TRUTH.

> *Unthinking respect for authority is the greatest enemy of truth.*
>
> *Albert Einstein*

6. Social Rule Book

Autistics often say that they haven't been given the social rule book, in defence of their apparent lack of social skills.

The missing social rule book actually sits within the limbic system of the brain, which is the area where mammalian evolution is recorded. However, the limbic system is being phased out as it comes to the end of its evolutionary assignment. Accordingly, the social rule book is becoming obsolete, as humanity increasingly opens to the beauty and power that's inherent in presenting the true self to the world. The masks of the persona are being discarded in a collective awakening that's revealing a radiant heart that beats as one.

The social rule book is not missing in autistics, it's just been upgraded to a guiding system that's more accurate, and true; an energetic system that's based on what's felt in the moment. This is part of the evolutionary shift from meeting *external* expectations to listening to the *internal* whisperings of our true nature. We're transforming from fear to love.

Everyone is partaking of this shift...in their own timeframe...which creates a *spectrum* of evolution. Autistics just happen to be the forerunners, but we all sit somewhere along that continuum. It benefits humanity as a whole when we uplift others, instead of putting them down, no matter how different they are. We all shine on the inside.

Autistics don't follow a social rule book, they follow their hearts. Their natural inclination is to be kind and loving, true to themselves, honest and full of integrity, non-judgmental and disliking of small-talk and gossip. These beautiful qualities are not skills to be developed, they're already sitting within the heart of humanity awaiting release.

Open your heart and throw away the rule book.

7. PATTERNS

I see patterns in everything except people and social dynamics.

(adult autistic)

That says it all.

Society is a melting pot of people's emotions and behaviours. Throw in a little more spice or a dash of passion, and mix in some bullying or a little jealousy...and the dynamics change. Every situation presents a different concoction of messy dynamics. Masks abound. Authenticity is rare. Game-playing is paramount. Appearances matter. It's a performance.

To beautifully sensitive autistics, this is a minefield. They see the beauty of Mother Nature in everything, so it's hard to fathom why neurotypicals are disguising their inner beauty in deference to what others around them think.

Mother nature's patterns are always beautiful, which is why they feel so good. Autistics choose to withdraw from social situations in order to stay aligned with the beautiful patterns of their inner *mother*-nature.

They're not antisocial, they just prefer to feel good.

Autistics have the reputation of being pattern-seekers. But aren't we all? Doesn't everyone see patterns? Perhaps in different situations or areas of interest...but patterns nonetheless.

A builder must be able to see a pattern in the structure before him; so too a dressmaker; an architect; an assembly-line worker; a mother, creating her day's timetable; any sports person playing any game; a musician; a doctor; a gardener.

Patterns are everywhere...in every pursuit. So why has the label of pattern-seekers been stuck to the foreheads of autistics, as if it's something out of the ordinary? They're just better at it than others. Celebrate...don't denigrate.

Perhaps it's time to start recognising those medical-model traits as patterns of higher ability.

8. INTERESTS

Once you are in a field of interest, references to that field appear all around you.

Ruth Wilson (2022), *The Jane Austen Remedy* (page 296)

What does this tell us?

Clearly, the imagination is magnetic. **[imagine it: see glossary]**

The more you place your attention on your imagined possibility, the more you attract the pieces that will eventually cobble together to become its physical manifestation, or that will prove its validity.

Autistics have the ability to tune into their imaginations and maintain complete focus on their topic of interest. They're incredibly magnetic as they pull together information that's innovative and unique.

The neurotypical world is increasingly recognising these gifts within autistics, and offering them employment opportunities that ultimately contribute to their company's progress and bottom line.

To say that autistics are fixated about their special interests, or that they drone on about them, is simply saying that their imaginations are open and flowing freely, and that their unique perspective has turned up a fascinating new pattern that piques their curiosity.

In seeing this as weird or different, you may be missing something very interesting...or profound.

9. DIAGNOSIS

DIA + GNOSIS = apart + knowledge; higher knowledge of spiritual things.

A diagnosis actually separates us from our higher spiritual knowledge. Our inner spiritual well-being is overlooked in a scientific diagnosis. Perhaps that's why the word diagnosis is defined as "scientific discrimination", where discrimination is "the making of distinctions, act of observing or marking a difference".

The word *diagnosis* screams *difference*. **[diagnosis: see glossary]**

From the moment you disclose your diagnosis, you're treated differently. Warning lights flash from the pitying eyes of those you've trusted to tell. No amount of convincing can un-disclose that fact or make you appear normal to them ever again.

I appreciate that neurotypicals may *think* they're treating us normally, but believe me, they're not.

I admit that my diagnosis *was* a relief because it made sense of my whole life, but it didn't change who I am intrinsically. It didn't make me more typical so I could fit into this world, instead of feeling like a misfit.

I remember, as a teenager, sitting alone in the lounge room watching the short, animated movie that graced our screens every Christmas. Narrated sensitively by Burl Ives, who also sang the songs, it was called *The Misfit*. Of course, it was about Rudolph, the red-nosed reindeer, who just didn't fit in because he was different. As I relate this past event, I can still feel myself sitting there with tears running down my face, because I knew exactly how Rudolph felt. I felt like that all the time. I, too, was a misfit, and it hurt terribly. Over the years, I've tried to source the movie, but to no avail. I hope that one day it does reappear, so I can view it with the awareness of my higher understanding.

It's taken years for me to reframe my life through a new autistic lens, and to gently cobble the shattered pieces of my crazy, ill-fitting identity into an eccentrically beautiful uniqueness.

It's taken years to realise that there's nothing wrong with me.

It's taken years for me to find my voice and to understand that no one owns my diagnosis but myself; that I'm the only one who's authorised to disclose my diagnosis...or not.

To neurotypicals, whether family and friends, medical professionals or those working in autism institutions, I ask you to be mindful when considering the disclosure of someone else's diagnosis. Stand in their shoes and ask yourself how that would make *you* feel. Better still, ask the autistic how they're feeling about their private medical information being made public.

According to the glossary definition, a diagnosis is the recognition of a disease from its symptoms, where a symptom means,

> *a physical or mental feature which is regarded as indicating a condition of disease, particularly such a feature that is apparent to the patient.*
>
> Online Oxford Dictionary

When the external physical and mental features described by a patient (symptoms) are the only basis for diagnosis, the root cause is overlooked. The root cause is always catalogued in your pre-programmed memory of past experiences, your limbic system. The root cause relates to your roots. The root cause is directly connected to your inner spiritual well-being; to the gentle whisperings of your protective inner *mother*-nature.

She's constantly communicating with you via your energetic feeling system. She'll try to get your attention if an experience is less than nurturing, initially by sending gentle messages...those little niggles that you'll often ignore. If you don't take action to change the external cause of this niggling *dis-ease*, her whispers will get louder as the energetic messages get stronger. Her most insistent message is the manifestation of *disease* in her body tissue...she's now feeling *ill-at-ease*.

Whenever she sends you a feeling of dis-ease, she's asking you to look at the immediate external experience that's triggering this feeling. She's guiding you to discern exactly what it is that you're allowing into your life that is not nurturing your spirit. She's inviting you to explore your inner feelings and the pre-programmed responses associated with them, right back to the original childhood experience. This is the root cause.

Once discerned, you can then take action to ensure that it doesn't reoccur. She simply wants you to feel good, and for her beautiful *mother*-nature to return to its state of ease.

Addressing the *symptoms* of any condition will never bring ease. Likewise, if you treat an autistic according to the *medical model* of expected characteristics, you're missing the truth. Only when you go within each individual and ascertain their unique spirit, will you be able to address their wellbeing, and discover that the root cause of autism is evolution.

The beauty that's inherent in the spectrum of evolution is that it does away with the *us* and *them* model of existence, as it accommodates individuality. When fitting-in is no longer a thing, universal acceptance will have been attained.

Diagnosis is superseded by gnosis on the evolutionary spectrum of spiritual wellbeing.

To autistics, I'd say, think carefully before disclosing your diagnosis to those who don't understand the big picture; unless you're truly ready to embrace your autism as the beautiful gift that it is.

10. IDENTITY

IDENTITY = SAME = FIT IN

The word identity stems from the Latin word *idem*, meaning *the same*.

Our identity is hewn from our upbringing; from our family, with whom we identify; and further to our friends and community.

Our identity ensures that we belong; that we are the same as; that we fit in.

But what happens when our individual uniqueness *doesn't* identify with the family, friends and community?

We're labelled different. Condemned to be an outsider. We just *don't* fit in.

We feel like misfits!

But why *must* we fit in?

How can someone else's way be better than our own natural inclination?

What purpose does it serve to deny our own true self in deference to what someone else dictates?

Fitting-in is the problem, because it creates resistance and dis-ease. It epitomises control. It begets unhappiness.

Autistics don't want to fit in. In fact, I'd hazard a guess that no one really wants to fit in; that they'd prefer to be themselves.

There needs to be a complete overhaul of the growing-up process, of crafting your identity, of societal dictates. *Fitting-in* needs to be thrown out and replaced with authenticity.

This is what autistics desire: to express their authentic selves to the world. After all, both the words *autistic* and *authentic* derive from the Latin for *self*.

Don't you want to be yourself? Don't we all?

How hard can that be, when it's already embedded within us? Why mess with nature?

Autistics are trying to show you a better way. An authentic way. Why not give it a try.

Let's all celebrate everyone's uniqueness.

11. EDUCATION

Autistic kids are coming through in their droves. They're multiplying. Why the alarm bells and flashing lights? Of course they are!

Our kids are changing...they're *evolving*...rapidly.

Isn't that obvious?

So why isn't the education system evolving...rapidly?

Instead of slapping another label on a brilliant kid's forehead, why not change the focus and look at yourselves!

The education system is failing the new generations...BIG TIME! There's nothing *wrong* with the *kids*. They're just part of the rapidly advancing generations of innovative, unique, quirky, out-of-the-box progeny.

What's *wrong* is the antiquated box they're being asked to inhabit. Let *them* lead *you*...and amazing things will happen.

This book says it all...

The Spark: A mother's story of nurturing, genius and autism
by Kristine Barnett

12. SEXUALITY

Autistics tend to show an inclination away from the normal binary gender preference. In many cases, they identify as pansexual, meaning that they are attracted to people regardless of their gender.

For me, this indicates a deeper, more spiritual connection as opposed to one that's based on sexuality. Autistics feel the energy of those around them, and use this as their guide when making connections.

Just like the geese on their flight path, autistics telepathically connect with those on the same, higher wavelength. These higher frequencies are always loving, kind, gentle and nurturing. Isn't this what we're all looking for in relationships? Isn't this how we all wish to feel?

The spiritual connection has nothing to do with chemistry or hormones driving our sexual urges. It has everything to do with how we *feel*.

Throughout evolution, as we come-up-over (overcome) our particular personal *agendas*, we rise above the divisions of *gender-*identity. And we connect lovingly with everyone, as we embody the higher frequencies.

Evolution is replacing chemistry with light. Again, it can be clearly seen that autistics are forerunners of evolution as they follow their inner light.

The prefix pan- means all, every, whole or all-inclusive.

Pan-sexual autistics love everyone. And one day, *every* human being will do the same.

How beautiful is that!

> "Autistics are just a petal on the flower of humanity.
>
> Samantha Craft aka Ciampi, Marcelle M.Ed

13. AN OBJECTIVE VIEWPOINT

Finally, I'd like to quote from an article written by well-renowned Australian journalist, Leigh Sales, about her experience in mentoring a group of autistic journalism students, who subsequently interviewed six celebrities for the ABC television series *The Assembly*.

Sales stated...

"It had a level of authenticity and sincerity that

I found very appealing."

"…they don't have any agendas."
"They're motivated by curiosity and they're very warm and non-threatening."

"I hope that perhaps the program debunks some myths about autism – falsehoods such as that people with autism don't feel emotion and don't have social skills," she said.
"I learned heaps from all of them…just about being spontaneous,"

"…they ask some very creative and original questions, and also some very direct questions…the autistic students asked very different questions to neurotypical interviewers!"

"…they made me feel that they had a lot of trust in me and they gave me a lot of warmth and affection."

Authenticity and sincerity, no agendas, curious, warm and non-threatening, spontaneous, socially adept and full of emotion and affection, creative and original as well as direct.

All of these qualities support my offerings in this book, and my belief that autistics are indeed the forerunners of evolution, showing the world how to be loving and kind and warm, whilst exhibiting curiosity, creativity and uniqueness.

©Marg Kinneen 2018 – 2025

Acknowledgements

My warmest hugs go to all Autistics, Aspies and Spectrum buddies. Your big task is to drag a sleepy world into the light. And you're doing so with grace and brilliance. Keep shining your beautiful lights.

To those few courageous souls who believed in me throughout this crazy process, I honour your bravery. You know who you are.

To my tech guru, the lovely Rachel Taylor, the universe sent me an angel-in-disguise when you graced my world. Many blessings.

My profound and most sincere gratitude to Helen Iles from Linellen Press, for your kindness, constant support and uplifting spirit. You're a wonder.

Grateful thanks to Ilsa Sharp for your thorough and informative Manuscript Assessment.

To my readers, thank you for your curiosity and stretchy imaginations that nudged you into the pages of my book. I hope your sense of wonder and delight has been piqued, and your powers of self-belief magnified.

Heartfelt thanks to all those who came before me in this great plan of evolution. What a mighty path you forged from light.

With special thanks to Baird. T. Spalding and his team of American archeologists, who courageously pushed the boundaries of their scientific understandings when, in the late nineteenth century, they encountered the masters of the Far East, who

> *"...proved conclusively that there is a Law that transcends death and that all humanity in its evolution is slowly moving forward to understand and use it...*
>
> *There is no question but these people have brought the Light through the long ages and they prove by their daily life and works that this Light does exist just as it did thousands of years ago."*

Spalding, Baird T. Life and Teaching of the Masters of the Far East (page 51)

Word-Magic Glossary

ADOLESCENCE = ADDLE ESSENCE = your beautiful, pure, true, original essence that's been festering angrily in your cells throughout childhood, having been suppressed in deference to adult control. Hence the fiery passion of adolescents, when puberty finally gives them a means of expression. The measure of fire and passion is directly related to the amount of suppression that was experienced during childhood. Your childhood suppression (depression = being pressed down or put under pressure) determines the chemical concoction in your cells, which directly relates to your behaviours. [see the definition for PASSION]

ADULT = A + DULT = DU eL Ty = DUALITY. The unified infant nature, upon acquiring the limbic mask through childhood programming, transforms into the dualistic adult persona, carrying masculine and feminine energies.

ALICE = ALETHIEA meaning truth, disclosure, unclosedness or unconcealment. Unclosedness = the state of being open, referring to the opening of the third eye. Disclosure = removing the skeletons from the closet; from their dark fiery cells. Unconcealment = revealing your true self in both spirit and body. The literal meaning of ALETHIEA is the state of not being hidden. When all the masks are removed, and all the individual personas are soothed, your light will no longer be dimmed, hiding behind cellular walls of fear. And finally,

ALETHIEA means the opposite of oblivion, forgetfulness or concealment, which are all definitions of Limbo, and its Limbic equivalent. So, when humanity comes out of hiding, truth will return to the human body; Alice will be in Wonderland. (Wikipedia)

ANCIENT = literally means *from before*.

ATMOSPHERE = ATOM + SPHERE = ADAM + SPHERE = ADAM'S + FEAR = MAN'S FEAR. Man creates the atmosphere, so choose your vibes mindfully. In the figurative sense, meaning the surrounding influences within your environment, it represents the atmospheric stimuli that triggers your nervous system...your system of *fear*! Man's fear triggers his own body's fearful feelings...and so the cycle rolls on and on.

ATROPHY = The literal meaning is without nourishment, or wasting away. But within this word, we can also see A TROPHY, meaning a prize of war, a sign of victory, a monument of an enemy's defeat. (Online Etymology) The under-functioning pineal is a monument to the defeat of inner sight, which is marked by atrophy. The pineal becomes a trophy of war instead of a portal of love. By association, it's the limbic system that becomes the trophy of war, the sign of victory, the monument of defeat over the inner sight of your original infant nature; an enemy's defeat, with the implied enemy being your original infant nature. [see the definition for PINEAL]

AUTISM 1 = AUTO + *-ism* = SELF *-ism* = the condition, state or practice of being oneself, where *self* means natural, not manufactured. Those with autism, therefore, prefer to stay true to their innate essence; to take guidance from within, rather than from the world around them, which wants to mold them to acceptable, social standards. They have attained the state of SELF-realisation.

AUTISM 2 = OUGHT + *-ism* = the condition, state or practice of withdrawing from the *ought-tos* and *shoulds*, the acceptable, social standards that are impressed upon individuals by the external world. An aversion to standards that are less than natural.

BACTERIA = BACK + TERRA (TERRAIN) = PAST TERROR/FEAR. As interior = inner terrain, and exterior = outer terrain, so bacteria = the past terrain/terra = past terror or fear.

BEAUTY = BE YOU *-ty*. Your true beauty shines from within when you're true to yourself. By connecting with your inner spirit, you open to the wonders of creation, and inspiration becomes your guide. *Be you*, and your unique gifts will express radiantly onto the canvas of your reality, and you cannot help but make your dreams come true.

BELIEF = BE + LIEF = BE + LOVE = BE LOVE. The *loving* steps you take toward the fulfillment of your dreams, are the evidence of your *belief*. You literally *love* your life into manifestation through the repetitive massaging of your creative thoughts, words and actions. Whatever you believe, becomes your reality.

BONE = BOW **N** = Nitrogenous *bow* = the *rainbow* of the visible light spectrum that forms our electro-magnetic atmosphere, which is comprised primarily of nitrogen. Bone, therefore, is created by the electro-magnetic cycling of fear and emotions through the body's neuro-endocrine systems, aligning with the surrounding electro-magnetic field. In short, bone is the dense remnant of the light-filled infant skeleton, its integrity having been chipped-away-at throughout childhood.

CALCIUM = CALLOUS *-ium* = that which is hardened; thick skinned; unfeeling or hardened in the mind. Calcium petrifies your connective tissue as it lays down the patriarchal laws and rules that become set in stone. You become thick skinned behind walls of emotion, as your gentle spirit cowers in fear from memories of hurt and pain; numbing your feelings becomes a survival technique.

CARBON = C + ARBOUR + N. Within the word carbon we find an *arbour*, or *flower garden*. This makes sense, as carbon is the organic building block of all matter, including our body tissue. It also has the property of holding smell, just like a flower. (MATTER = EM Attar/perfume = electromagnetic floral perfume/aroma) (CHROMOSOME = Carbon aroma body) The aromatic fingerprint of all matter is based within the flower garden of carbon. [see the definition for MATTER, CHEMISTRY & CHROMOSOME]

CHEMISTRY = C EMystery = CARBON MYSTERY. Carbon, your body's organic building block, carries your aromatic signature in the form of smell. Your body matter is carbon infused with smell, courtesy of the chemistry in your cells. Your smelly memories of the past that are inherent in your chemistry, veil your organic building block, carbon, and create a mystery of her pure nature. Carbon lies hidden behind smell/chemistry. [see the definition for CHROMOSOME, MATTER & OLFACTORY]

CHRISTMAS TREE = PINE tree adorned with light = PINEAL or open third eye. The third eye is represented by an eye within a triangle, surrounded by radiant light. The triangle = the Christmas pine. The radiant light = the Christmas lights strung around the tree, and glowing in your brain's corona radiata (radiant crown/halo). The eye = the higher vision when illusion (ill vision) is fully evolved. TRIANGLE = DELTA = the fourth letter of the Greek alphabet and the infant's brainwave frequency of intuition, imagination and regeneration. The open third eye image is a stylized representation of the intact ovum, with its all-seeing connection to the infinite field of creation, and its surrounding corona radiata (radiant crown). When your body reconnects its central canal to your ovum, your Christ light is born. Joy to the world.

CHROMOSOME = C + AROMA + SOMA = CARBON AROMA BODY. Your DNA is packaged in the form of chromosomes. Each chromosome is a body of carbon carrying aroma or smell, which is the memory that's been passed down by your ancestors. Your chromosomes are your aromatic fingerprint. [see the definition for CHEMISTRY & KARMA]

CIRCUMPUNCT = CIRCLE + POINT, being that its symbol is a dot within a circle. Also, CIRCUIT + POINT, being the point of power or power point...the infinite source of power and first cause of creation. Also CYCLE + POINT, being the point through which the cyclical generations of evolution enter in round after round of life times.

COGNITION = CO-IGNITION. Cognition or thought is simply the firing together of nerve cells along a particular path, after an external experience has impacted upon the body. With constant revisiting of that same pathway, the thought strengthens and becomes a habit. Thoughts are stored in proteins at the points of co-ignition, so the more frequently you revisit the thought, the stronger the protein gets, the stronger the habit becomes. To break your habits, change your thoughts. [see the definition for PROTEIN]

COLLAGEN = COLLEGE + N = NITROGENOUS COLLEGE. Calcium is passing on its rigid patriarchal teachings (DNA) within the nitrogenous college of your physical body. Tradition sets the masculine firmly within the walls of your corporeal temple of learning.

CONNECT = CON + NECTAR = WITH NECTAR. Nectar is a delicious drink or the sweet liquid in flowers, and is seen to be the legendary drink of the gods that bestows immortality. **NECTAR** = NEK + TAR = DEATH + OVERCOMING = OVERCOMING DEATH. Your connective tissue carries your essence through the generations, slowly refining and purifying its measure of sweetness, whilst slowing down the aging process. It's your vehicle for evolution.

CONNECTIVE TISSUE = TISSUE that CONNECTS = 'TIS YOU that CONNECTS. You connect with everyone and everything in the universe through your connective tissue. Your connective tissue is the threads of your life; the yarns that you've collected; the fairytale that narrates your character and pulses through your body tissue.

CORONA RADIATA = RADIANT CROWN. A corona can also be a luminous halo or ring of light, as depicted around the heads of saints. It's also a garland of flowers and an ornament of refined gold.

DEFENCE = D FENCE = DELTA FENCE. The delta brainwave patterns are predominant in the INFANT, which is your true design, beginning with the embryonic infant stage and continuing through birth, but slowly diminishing during childhood. The pure, innocent and happy properties of DELTA, which radiate from within, are intuition, imagination and regeneration. But with the onslaught of the external world, the INFANT takes refuge behind a chemical wall of fear in order to protect its beautiful treasure and survive in this world. It's building FENCES...the DELTA (D) FENCES...it's DEFENCES. Our true design has gone into hiding due to the messages of fear that are being impressed upon it by the well-meaning adult world. [see the definition for INFANT & INFINITE]

DELIGHT = D LIGHT = DELTA LIGHT of infants...little bundles of joy. Delta brings intuition, imagination and regeneration. The Garden of Eden is the garden of delights...your flourishing ovum of roses, informing your body with the highest essence. [see the definition for DELTA, JOY & FRUIT]

DELTA: Delta brainwave patterns are inherent in the INFANT, from embryonic infancy, through birth and into childhood, wherein they slowly diminish. The properties of delta brainwave patterns are intuition, imagination and regeneration, due to the infant's inherent connection to the universal power source...the treasury of love, dreams and youth...that exists on the highest light frequencies of self-belief. The higher power creates the brightest light. INTUITION = INNER TUITION = loving guidance from the higher power (self-belief) within. IMAGINATION: IMAGINE IT = MAGNET = the power to attract unlimited possibilities. REGENERATION = lovingly nurturing the body tissue with the ease that's inherent in the highest light frequencies, thus recreating youth in every moment. Your body tissue delights in your self-belief. Delta provides the radiance of the pregnant mother...the little light of her baby...you. Delta is your little light...within. (This little light of mine...I'm going to let it shine)

DEPRESSED = DE + PRESSED = DOWN PRESSED. When your true essence or nature is pushed down beneath a mask of acceptability or expectation, your spirits feel depressed...because they literally are being pressed down; they're under pressure. They're sad because they are forced to hide behind a false face of somebody else's creation. They're feeling *not good enough*, because they are not approved of and therefore cannot be expressed.

DESIRE = D's IRE = DELTA's IRE. The delta brainwave patterns are those of the original, pure infant, through whom the infinite, universal font of love and creativity flows; intuition, imagination and regeneration flourish within the little bundle of joy. However, the human persona very soon settles its mask of habit within the infant's tissue, thus concealing its true nature, and provoking IRE...anger, passion, drive, madness, gadfly (that which irritates). Interestingly, estrus (the phase of the cycle when the female is sexually receptive and capable of conceiving) also means passion, mad impulse, frenzy and gadfly. The infant is conceived through DESIRE; through your body's mad and passionate impulse to rid the irritating imposter, DNA, and express her true nature. She's angrily throwing off the fiery false mask of habit through the cyclical cleaning-frenzy of childbirth…otherwise known as evolution. [see the definition for EMBRYO and PHOENIX]

DEVIL = DIABLO *f.* Fr. DIABLE = DOUBLE (D + uble). The devil, or slithering reptilian from the Garden of Eden parable, represents the dualistic nature of your DNA that's handed down through the family tree in the form of the masculine and feminine characteristics that create your human persona. The family tree is the tree of the knowledge of good and evil; feminine and masculine; high-vibes and low-vibes.

DIAGNOSIS = DIA + GNOSIS = apart + knowledge; higher knowledge of spiritual things. A diagnosis actually separates us from our higher spiritual knowledge. As the recognition of a disease from its symptoms, a diagnosis is clearly an externally-derived prediction that doesn't address the root cause. Our inner spiritual wellbeing is overlooked in a scientific diagnosis. Perhaps that's why the word diagnosis is defined as scientific discrimination, where discrimination is the making of distinctions, act of observing or marking a difference. The word diagnosis screams difference. (Online Etymology Dictionary)

DISEASE = DIS + EASE. When you're experiencing unpleasant situations or thoughts, you send heavy vibrations into your body tissue by way of dense chemical messages. It's these heavy vibes that create dis-ease in the very fabric of your being. If you don't listen to how your body is feeling, and change your habits, the dis-ease will create disease. Your body is imploring you to start nurturing yourself.

DUALITY = DU-L-T = DULT = ADULT. The establishing of masculine and feminine energies within the cells leads to the sexual identification of the individual at adulthood. The adult epitomises duality, just as the purity within the infant epitomises unity.

DURATION ƒ Latin *durus* meaning hard, refers to the length of time something continues, providing hard evidence of its existence. [see the definition for ENDURANCE]

EMBRYO = EMBER + RIO = FIRE & ASH + CYCLE. The embryo is the proverbial phoenix that continually rises from the ashes to ashes, dust to dust cycles of evolution. With each new cycle, DNA is revised, along with family beliefs and behaviours, thus providing an increasingly youthful evolutionary legacy. [see the definition for PHOENIX]

EMOTION 1 = E.M. OCEAN = **E**LECTRO **M**AGNETIC OCEAN = **E**LECTRO **M**AGNETIC FIELD. Emotion is the salty electro-chemical ocean that's stored within your cells, courtesy of your neuro-endocrine system, and subsequently contributing to the collective field of emotion that's otherwise known as the electro-magnetic field. The electro-magnetic field, therefore, is a field containing the full spectrum of emotion. Emotion is the residue of ancestral/family baggage that goes out into the electro-magnetic field to explore its beliefs and behaviours in relation to others.

EMOTION 2 = E + MOTION, meaning an outward stirring or agitation...a mental or physical tossing to and fro...casts *emotion* in an unfavourable and dualistic hue. With closer inspection we find that the whole spectrum of emotions ranges from the densest to the least dense vibrations...although *dense* nonetheless. However, when we reach the evolutionary pinnacle, having overcome the whole spectrum of emotions, we leave the electro-magnetic field behind and enter a purely magnetic field. The goal of evolution is accomplished. [see the definition for EVOLUTION]

ENCLOSE = CLOSE/SHUT IN or IN THE CLOSET or IN SECRET or CONFINE or CONCLUDE and FENCE IN. [see the definition for DEFENCE]

ENDOCRINE = ENDO + CRINE = IN-HOUSE + SECRETS = FAMILY SECRETS = FAMILY SECRETIONS = CHEMISTRY. The endocrine glands and organs make and release an internal chemical secretion called hormones. The literal translation of endocrine is *secreting internally*, where ENDO is derived from the Latin IN + DOM (house), and CRINE is the Latinized form of the Greek *krinein* meaning to separate, distinguish or discriminate. ENDOCRINE = stored family patterns of judgement that distinguish themselves from others, setting themselves apart from others, and marking the points of difference between themselves and others. The endocrine glands, following the fear-filled script of the nervous system, contain the chemical bible of family judgement and expectation...their religious and cultural beliefs that distinguish them from others.

ENDURANCE = f Latin *durus* meaning hard, is the art of tolerating or undergoing suffering; to harden (the body) against prevailing conditions through suffering (resulting in one's passion). The individual's original infant design is hardened, in order to endure the prevailing external conditions that are impacted upon it. Suffer = *f.* L *suf-* + *fer* = to carry under. Passion = suffering. The infant design is pushed into hiding beneath the hard outer shell of protection, thus transforming into the fiery inner passion that lay deep in your limbic system as memory and emotion. [see the definition for DURATION]

ENERGY = INNER G = INNER GAMMA = the inner power of creation. Our creative energy is measured by our connection to gamma. Gamma has the highest frequency, and therefore carries the highest measure of light and energy. Gamma resides in the gamma field at all times. But as we move into lower frequencies, we literally reduce the number of times (frequency) that we tune into the gamma field.". We lose power and dim our lights. Our power of creation diminishes. (See definition for GENIUS, GENERATION, GAMMA, GENE and MITOCHONDRIA)

EVOLUTION 1 = EF + ILLUSION = OUT OF ILLUSION. (With an *f* and *v* interchange in language, *ev* becomes *ef*, both relating to *ex*) Evolution is the process through which humanity is rising out of the illusion of low self-belief by reconnecting to the high-powered inner guidance system that was inherent in its infant form. Generation after generation is seeing a progressive refinement of the essence of genius; or indeed, a slowing down of its decline within each individual's life time. [see the definition for ILLUSION and EVOLUTION 3]

EVOLUTION 2 = EF + ILLUSION = EF + ILL VISION. (with a *u* and *v* interchange in language, ill*u*sion becomes ill *v*ision) Evolution is the process wherein humanity is rising out of the lower light frequencies of the visible light spectrum, into the higher light frequencies of the fully open imagination (gamma). Humanity is opening to a higher vision.

EVOLUTION 3 = EF + OLU *-tion* = the action of coming OUT OF OIL. With a *u* and *v* interchange in language, OLU becomes OLV or *OLIVE*, which derives from the Latin for *oil*. And what is it that comes out of oil? Essence, of course. So, evolution is the long-running active process of expressing a progressively more purified essence or smell, which is giving rise to the higher vision and more fully open imagination of creative genius...within everyone! [see the definition for EVOLUTION 1 & 2]

EXPECTATION = EX + SPECTRE *-ation* = OUT + TO LOOK = the state of LOOKING OUT, with the literal meaning being that of *waiting*. So, the WEIGHT of expectation is the WAIT for external approval. Expectations determine your OUTLOOK on life, your perception that casts its controlling net over your reality.

FACT: ƒ Med Latin, meaning a thing done, something that has actually occurred, a thing known to be true. A fact is something that we already know, which is why science deals with facts, because the word science means state or fact of knowing; what is known; knowledge; ƒ Latin *scientia*. (Online Etymology Dictionary)

FATE = ƒ Latin *fari*, meaning to speak, tell or say. Understood to be a prophetic declaration of what must be, fate clearly bestows power upon your words as they craft your reality. Your fate, therefore, is that which is ordained...by you! (Online Etymology Dictionary)

FIRE = EF + IRE = EX + ANGER/INFLAMMATION = from anger or from within flames (flames from within). Male/masculine is represented by fire; the inner flames of anger.

FLOWER 1 = FULL HOUR. When a plant flowers, it's reached the fullest expression of itself. It's achieved its finest hour. The word *flower* is also figuratively used to indicate innocence, purity, elite or something that's at its prime; it's first or finest embodiment. Through the refinement of body essence, evolution is fostering the blossoming of humanity into its fullest expression of purity, beauty and youth. The full hour of evolution = the flower of evolution, whose essence is refining within your cells = leading to the purest essence of roses. [see the definition for PRINCE CHARMING]

FLOWER 2 = FLOW -er = that which FLOWS. That which flows within your body is both your blood flowing through blood vessels, and your energy flowing through your meridians. They are intimately connected as they channel your very essence through your body tissue. And when they reach their fullest hour, your meridians will be operating at their fullest power, as the essence of roses flows through your body.

FREEMASON = FREE + WORKER IN STONE. It's my belief that Freemasons know that evolution is the process of rising out of fear and into love, both in spirit and body. When this is achieved, your body tissue will no longer bear the signs of petrification, ossification, mineralisation and atrophy that accompany fear-filled beliefs. Your tissue will be as youthful and soft as it was in your infancy. You'll be free of the stony demeanour that comes with limitation and aging.

FRUIT = from the Latin *fructus*, meaning an enjoyment, delight, to enjoy. Originally stemming from the products of the soil, and extending to income, revenue and profit. (Online Etymology) The fruits of your labour are the outcomes or consequences you enjoy. And of course, your offspring (the fruit of your loins) are little bundles of joy. Hence, the level of joy that courses through your body tissue is dependent upon the seeds you sow...your thoughts and beliefs. Joy and youthful body tissue are the fruit of your beliefs. DELIGHT = D LIGHT = DELTA LIGHT of infants...little bundles of joy. [see the definition for DELTA, INFANT and JOY]

FUNCTION = EF + UNCTION = OUT OF or FROM + OIL = that which comes out of, or from, oil = ESSENCE. Your olfactory system of smell manufactures your essential oils that determine the exact function of each and every cell. Therefore, the function of a cell is encoded in smell. And every function in your body is determined by smell. Smell carries instructions or destructions, depending on its composition. The cell's functions are constantly evolving. [see the definition for CHROMOSOME, MATTER, OLFACTORY and EVOLUTION]

GAMMA: The source or power of creation within each and every human being, which flows through everything. The source of GENIUS. The magnetic power that generates the highest light frequency and radiant youth when at its peak. Gamma is the highest light frequency in the electromagnetic spectrum, inspiring the most energy and light.

GARDEN 1 = GUARD **N** = NITROGENOUS PROTECTOR. The perfume of your body chemistry is designed by your limbic system's nitrogenous genetic memory as it protects and defends your organic body matter from perceived external threats. The plant and animal kingdom use these scented messages to navigate the world around them. So too does humanity, albeit with less conscious awareness.

GARDEN 2 = **G** ARDOUR **N** = EARTH BURNING NITROGEN. **G**, from Greek *ge* meaning earth. Ardour is the heat of passion or desire, inflammation, fire and zealous eagerness. **N** = nitrogenous DNA. The garden in your cells is the perfume exuded by your body chemistry, when infused with the dense emotions of unrequited dreams, and the inability to express your truth, due to external pressures. Your cells are filled with a fiery passion and anger that incites inflammation; the burning sacrifice of your true essence upon the altar of childhood, leaving a putrefied concoction of hormonal instinct (inner stink), and a thorny barrier of protection.

GARDEN 3 = **G** ARDOUR **N** = GAMMA + WHOLEHEARTEDNESS + NITROGEN. Gamma is the God vibe containing the highest measures of light and energy. When the garden in your cells fully evolves, it will exude the wholehearted goodness and love and sweet essence of roses that epitomizes the higher power called God.

GENDER = AGENDA = that which needs to be done. The characteristics ascribed to each gender is accompanied by the hidden agenda of expectation that is currently accepted as normal at that time. Evolution has seen those expectations altered from generation to generation. As the agenda changes, so too does the gender.

GENE 1 According to Wikipedia, a gene is a region of your DNA that determines function. The genes' code is in the form of smell, which sets up your essence in your cells. [see the definition for CHROMOSOME & FUNCTION]

GENE 2 = GENIE. Your genes embody the spirit of genius that magically creates your body and your world...your very own magic *genie*. Your spirit of genius is smell. [see the definition for GENE 1 & CHROMOSOME]

GENE 3 = G *-ine* = *relating to* GAMMA (G). Your energy is the *inner G* or gamma within your cells...your power of creation. Creation is generation. [see the definition for GAMMA, GENERATION and GENIUS]

GENERATION 1 = GENE ORATION = GENE EXPRESSION = *how* the genetic information within DNA takes the form of your body matter... how it expresses through, and shapes, your physical body. Your genetic information *generates/creates* your body matter. Scientific evidence supports the fact that genetic information is shaped by your emotions, which are in turn shaped by your beliefs. BELIEF is the power of generation or creation therefore belief is your inner genius. Ultimately your beliefs determine the genetic information that guides the creation of your body. Your beliefs are your whispering inner *genie*. Higher self-belief generates a higher vision, and a more magical genie. Ultimately your beliefs determine the genetic information that guides the creation of your body.

GENERATION 2 = **G** NARRATION = GAMMA NARRATION = GAMMA STORY = GAMMA STORE in your cells. Therefore, the INNER GAMMA STORE = INNER G STORE = ENERGY STORE. Your genetic information is the *family **story*** that's been passed onto you from your ancestors via DNA. These *familiar beliefs* are the *energy store* within your cells, which generates/creates your physical body (see previous definition). Telling yourself a loving story and filling your cells with creative energy, will generate a youthful body. GAMMA is your generative/creative power or GENIUS, otherwise known as BELIEF. You become your own tutelary spirit or guardian when you tell yourself a loving story, and reconnect with your inner genius. Each new generation holds a more loving store of energy that generates a more youthful body.

GENERATOR = **G** NARRATOR = (INNER **G**)/ENERGY NARRATOR = MITOCHONDRIA. The story teller of your beliefs. The creator of power.

GENIUS 1: A tutelary spirit or guardian that watches over you, otherwise known as a *genie*. Further to this, the original meaning was *generative power*...the power of creation...creativity. (Online Etymology) Genius is your power of creation. Your measure of genius depends upon your connection to the infinite universal field of creation, your higher belief system; your higher power. In referring to the definition for INFANT, you'll see that it's your creative genius that powers your body during infancy, thus facilitating youth, which is evident through the radiance of your body tissue due to its high frequency of light. INFANCY = GENIUS. Unfortunately, as you lose touch with your loving inner genie during childhood, your creative power diminishes, and your radiant light dims. We call this diminished expression of light within your body tissue, aging. The exalted natural ability that's associated with the word *genius* is inherent within the state of infancy. [see the definition for INFANT, INFANCY and GENERATION 1 & 2]

GENIUS 2 = **G** -*ness* = GAMMA -*ness*. The measure of gamma into which you tune, determining your expression of light and energy. Your higher belief system or higher power. (see the definition for GENERATION and GAMMA).

GENOME = **G** + NOME = GNOME. Your genome is the full set of DNA instructions found in your cells. It provides the blueprint for the development and function of your earthly body, within which you dwell for your lifetime. Interestingly, a GNOME is defined as a legendary dwarfish creature supposed to guard the earth's treasures underground. (Oxford book of Languages) Clearly the seven dwarfs are representative of your endocrine hormones that provide the instructions to your cells, the family patterns of emotion, which buries your true essence in the earthly tomb of your *bony* skeleton, beneath its inorganic mineral content. Your treasured light has gone underground.

GEOMETRY 1 = GEO + METER or MEASURE = EARTH MEASURE. The measure of any object is the limits or boundaries that define its finite appearance, as formed by the light photons in the electromagnetic field. Those limits and boundaries take the form of points, lines, curves and surfaces. Those limits and boundaries are dictated by your emotions, which are determined by the stored memories in your limbic brain. The geometry of your reality is designed by your memory and emotion...your limbic system. **The function of your limbic system is to create your reality.**

GEOMETRY 2 = GEO + METER or MEASURE or MOON (from the same Latin derivative, the moon being the star that measures time) = EARTH + MOON = EARTH MEASURE = geometry is your earthly limits/boundaries that form your reality = SPACE/TIME. The geometry of space/time derives from the stored memories of your limbic system.

HEART: A centre or nucleus. The cell nucleus of every cell in your body is the seat of your emotions, courage, desire, spirit and will; the repository of your mind, intellect and memory. It's here that your life fairy story finds its beginning, middle and end, written in the evolving energy of belief. When you have a change of heart, you change the information in your cells, and as a result, your whole body transforms. Evolution *is* a change of heart.

HEART 2 = HER + ART = HER ART. The feminine expressing her creative flair upon your body tissue, which is the living art gallery of your life's work.

HELL = HADES = THE UNDERWORLD. The word hell literally means to cover or conceal (Oxford Concise Dictionary). And Hades means the unseen one (Wikipedia). The Underworld is clearly that which is hidden.

HOLY = WHOLLY. Therefore, your HOLY SCRIPTURES = WHOLLY + THE WORD OF GOD = FULL BELIEF. When you *religiously believe* in yourself, your cells will carry the highest energies; the highest light frequencies of a fully open imagination. You will be the embodiment of the creative, ever-loving, ever-nurturing *mother*-nature.

HOLY SPIRIT = WHOLE SPIRIT = FULL INSPIRATION or essence of truth and integrity. When the Holy Spirit flows through your body, every fibre of your being will be filled with your unique essence, manifesting through physical integrity as youth, your true design. She is the essence of forgiveness that cleanses your cells of the past, so you can lovingly receive her gift of the present.

HORMONE 1 = HER AWE MOAN, where awe is defined as wonder. Wonder leaps from the infant and child, and weaves its magic within the youthful threads of their body tissue. Wonder is inherent in the delta brainwave patterns of infancy and childhood. Unfortunately, it's the infant's wonder that loses the battle against the external world with its tight strictures, and becomes buried deep beneath the accepted patterns of growing up. Wonder is usurped by the accepted norms of adulthood...the rigid rules of expectation. However, wonder doesn't disappear. She slips into hiding during childhood, awaiting release in the form of hormones at the time of puberty...the onset of adulthood. The infant has been adulterated, and she angrily bemoans her fate through the chemical echo within her cells. The pure innocence of the infant is wearing a mask of passion as her true essence wallows in a fiery pit of anger called desire...her unexpressed heartfelt dreams. Hormones are your true essence in captivity...your dreams that wish to be set free...your passion...your suffering.

HORMONE 2: The word hormone literally means *impetus*, *drive* or *to set in motion*. Accordingly, hormones take on a regulatory function, controlling the stimulation or inhibition of bodily processes. Hormones are the chemical soldiers that maintain obedience and routine within the nervous and endocrine systems. Hormones dictate through fear and family obligation.

HUMAN = HUMOUR **N** = NITROGENOUS HUMOURS. The four cardinal humours are blood, phlegm, choler and melancholy, which are associated with particular organs. The mixture of these humours determines a person's mental, emotional and physical qualities, resulting in their disposition and complexion.

HUMAN RACE = a RACE is a CANAL or CHANNEL. The human *race* is the *channel* through which the nitrogenous humours reign. It centres around the spinal *canal*, running from base to crown, informing every system of the human body. Every individual's human *race* is unique to them, and it contributes to the collective energies of humanity within the electromagnetic light field.

HYALINE CANAL = HIGHER LINE CHANNEL. When open and connected, in your infancy, your hyaline canal channels the higher power from your ovum, through the centre of your spinal canal, to your third eye. It is your direct line to God (or whatever term you use for the creator).

ILLUSION = ILL VISION. With a *u* and *v* interchange in language, we find that ill*u*sion becomes ill *v*ision. When disconnected from the universal field of creation, you fall into the lower vibrations of fear and emotion, thus holding limited beliefs about yourself and your life. Lower self-belief creates a lower, or ill, vision for yourself, which is an illusion established by the fear-filled external world. [see the definition for EVOLUTION]

IMAGINATION = Where your imaginings reside; the gathering of your magnetic essence; your field of attraction; the source of creation. [see the definition for IMAGINE IT and MAGNET]

IMAGINE IT = M-A-G-NET = MAGNET. Your imagination is the most powerful force in the universe. It's magnetic, and attracts your reality to you. [see the definition for MAGNET, IMAGINATION and PRINCE CHARMING]

IMMUNE SYSTEM 1 = INNER MOON SYSTEM. The immune system is the mothering system within your body. And just like a mother, she angrily swoops in to protect your body tissue whenever it's under threat...inflammation is her first line of defence. She works with your limbic system building fiery energetic walls of defence. The moon is the star that measures time, therefore the immune system is the system within your body that also measures time. In alignment with the limbic system, she creates the mask of mortality...contributing to evolution by swaddling your wounded body tissue and carrying it through time.

IMMUNE SYSTEM 2 = *In-* + MOON SYSTEM = OPPOSITE OF MOON SYSTEM = SUN SYSTEM or SOLAR SYSTEM. The *angry* mother, intent on protecting her child, is analogous to the *inflamed* matter of the immune response...*fiery* in defence...the wounded feminine in her masculine mask.

IMMUNITY = *Im-* + UNITY = INNER UNITY. When you reconnect with your *inner* power source, at the base of your spine, your 7 chakras will be reunited in a stream of pure white light, bringing full immunity to your body. You'll have reconnected with your source of creative mother energy, soothing and nurturing your body tissue, thus reinstating your original design of youth.

INFANCY = IN + FANCY = INNER FANCY = INNER FANTASY. From embryonic infancy through to childhood, delta brainwave patterns predominate, facilitating imagination, intuition and regeneration. The creative imagination, inner fantasy or dream-world is vivid. Far from being illusory, this is the state of higher visions and unlimited possibilities; the field of creation, where creativity abounds. [see the definition for INFANT]

INFANT 1 = INFINITE. From embryonic infancy you're connected to the infinite universal field of creation, and the highest light frequencies. This field of unlimited possibilities is the source of the imagination, all knowledge and pure love. In scientific terms, it's sourced through the pluripotent embryonic stem cells, which hold the power to differentiate into unlimited possibilities. In spiritual terms, it's the field of creation, the field of grace where love reigns supreme and inner guidance is engaged. Two different perspectives...one phenomenon.

INFANT 2 = INNER FONT = INNER FOUNTAIN. With the inherent property of *youth*, it can be seen that the INFANT = INNER FOUNTAIN OF YOUTH. As a place of rising, a fountain or font can also be called a wellspring or source. The INFANT or INFINITE state of being is the deep inner well of creation, where the imagination flows with unlimited possibilities. [see the definition for INFANT and DELTA]

INFANT 3 = Latin *in-* <u>the opposite of</u> + *fari* <u>speak</u> = <u>not speak</u>. The infant has no audible voice, for its communication occurs via light frequencies...telepathy. It is ineffable...its essence being beyond expression, too great for words because words require sound, which is of a lower light frequency. The infant carries the ineffable word of God (creator), the highest light frequency of love.

INFINITE = INFANT. The word *infinite* means eternal, limitless, endless, boundless, with no end or finish. (Oxford Etymology) Embryonic infancy is the boundless state that is open to unlimited possibilities (think stem cells), which occurs before the onset of the limbic system. Interestingly the word *limbic* means border, serving to reinforce the understanding that the limbic system protects the infant genius-state by creating chemical borders behind which it is masked. INFINITE = not finite = no borders = NO LIMBIC. [see the definition for LIMBIC]

INFORMATION = INNER FORMATION. The information coursing through your body systems creates the inner formation of your body. The electrical messages of your nervous system inform your brain and create the chemical messages within your cells, which in turn determine your body structure. By changing your beliefs (information), you alter the structure of your body.

INSPIRATION = the state or condition of your INNER SPIRIT. When inspiration strikes, your inner spirit becomes animated with creativity.

INSTINCT = INNER STINK = A strong offensive odour from within. The cells filled with hormones, the chemical concoction, or smelly prisoners that are based upon memories from childhood as you responded to the external world. You smelled fear, and stored it in your cellular memory, in readiness for its release during puberty. Instinct comes from the Latin for *in* + *prick*. I'm hearing strains of Sleeping Beauty! After being pricked by the sharp reality of the external adult world, you lose touch with your inner beauty as you fall headlong into the illusory sleep of chemistry within your cells. Instinct is a narcotic.

INTELLIGENCE 1 = IN + TELL + **G** -*ness* = INNER TALE of GENUS = the story in your cells = the family tales you tell yourself that become your repeated beliefs, your proteins and therefore your cellular information. The words *tell* and *tale* defer to each other in the dictionary of etymology, making them interchangeable.

INTELLIGENCE 2 = IN + TELL + **G** -*ness* = INNER TALE OF GAMMA. Gamma is the highest light frequency, carrying the highest measure of light and energy. Gamma is the higher power, or field of higher information, that we're plugging back into through our evolutionary climb. Our measure of gamma indicates our level of empowerment and brilliance, or enlightenment. Effectively, gamma is God, reflecting our religious beliefs...the beliefs that we practice religiously. Loving beliefs, as whispered by our motherly mitochondria, fill us with light bringing wholeness to our body tissues. Figuratively, we become one with God...our higher truth.

INTUITION = INNER TUITION. Your loving higher guidance that watches over and protects you from within, through your energetic communication system called feelings. It's otherwise known as *self*-belief, which is always ready to inspire your creative genius. Intuition and genius are synonymous. You can recognise your intuition because it's always loving...and it always believes in you. Intuition powers all your *creative* endeavours. Pinocchio's Turquoise Fairy is representative of your intuition.

JESUS = CHRIST JESUS = X JESUS = EXEGESIS. Exegesis means to explain...to make plain; to interpret...give a stylistic representation of a creative work. Your exegesis is your creative expression or the out-pressing of your uniqueness; your individual gene-expression; your DNA. Jesus = YHWH = TETRA-GRAMMATON = FOUR LETTERS = just like DNA's four bases, ACGT.

JOY = Pleasure, delight, happiness, bliss. This becomes interesting when we align it with the word *fruit*, which not only alludes to offspring or produce, but also derives from the Latin *fructus*, meaning joy/enjoy. The fruit of your beliefs is evident in your body tissue. The joy you sow into the cells of your earthly nitrogenous body matter, bears the fruit of youth in equal measure. [see the definition for FRUIT, DELIGHT & DELTA]

KARMA = C + ARMOUR = CARBON ARMOUR. The limbic brain is your physical defence system, which sends chemical soldiers into your cells, creating the emotional armour that protects you from perceived threat. This defensive, emotional chemistry sits within your organic building block, carbon, becoming your CARBON ARMOUR or KARMA. In Buddhism, the word karma means "the sum of a person's actions in this and previous states of existence, viewed as deciding their fate in future existences". (Oxford Reference) Clearly, the limbic brain stores the memory of a person's previous actions, which then determines the chemical fate of their cells and their subsequent emotional behaviour. Thankfully, the limbic brain is progressively discarding it's suit of emotional armour throughout the evolutionary process, which will also put an end to karma. [see the definition for CHEMISTRY and CHROMOSOME]

KUNDALINI = C + UNDER -*lini* = pertaining to the pure CARBON that lay hidden behind the smelly chemistry within your cells, awaiting resurrection through the process of evolution. Your kundalini rises through the full expression of your true essence within your body tissue, culminating in youth. She is your divine feminine *mother*-nature holding a beautiful bouquet of roses. [see the definition for LATENT]

LATENT = LIE HID = LYING DORMANT = SLEEPING. Your pure spiritual energy is in existence but not yet made manifest; it's concealed; in the closet, waiting for you to open the door so it can step into the light of day.

LIFE = ENERGY. Just as the life of a battery is the energy within it, so too is your *life* measured by the *energy* within your body. Your beliefs determine the energy in your cells, which manifests as your life; your reality. If you don't like your life, you can change your beliefs and create a new one.

LIMBIC = BORDER (LIMBO). The limbic system is the memory and emotion centre of the brain that's established during childhood. It sets up chemical borders, behind which the infant (genius) takes refuge from the onslaught of external stimuli...as your self-belief is eroded. This fear-based defence system is building protective fences for the beautiful delta brainwave state of infancy...DEFENCES = D FENCES = DELTA FENCES. The limbic system is constantly determining your level of safety, and allocating emotional values to each external stimulus that you experience, based on its bank of past memories, thus determining your chemical reactivity. The fact that it's a RE-action tells you that you're re-enacting a past experience that occurred initially in your childhood. [see the definition for DEFENCE and DELTA]

LIMBO: LIMBO = LIMBIC. Limbo is a condition of neglect or oblivion; a place of confinement and forgetfulness. (Online Etymology) The limbic brain represents Limbo, where your beautiful infant essence is confined and neglected; where your first and fondest memory is cowering in fear; where your soothing *mother*-nature is wounded, disguised as your fiery masculine or reptilian nature. The confinement and neglect are transported into your cells, being the prisons where you serve your sentence of death...mortality.

LUCIFER = LIGHT BRINGER = TORCH BEARER. Lucifer is known to be the spirit of genius or the spirit of evil. He's analogous to Satan, the Devil, Beelzebub, the planet Venus, the Morning Star, the god Attar and Phosphorus. Lucifer is said to be the son of Aurora (Dawn), which relates etymologically to Ostara, Eostre and Easter, all pertaining to a rising or resurrection. And Lucifer has been represented as a winged child, pouring light from a jar. (Wikipedia) Lucifer is quite clearly indicative of the light-bringing, angelic infant who falls from the heavenly heights of the highest light frequencies through its childhood journey into adulthood, gathering emotional dis-ease in its cells.

MAGIC = IMAGE -*ic*. Your imagination is all powerful. In fact, it's the most powerful force in the universe. IMAGINE IT = MAGNET. Whatever image you visualise, with focused and loving attention, will manifest into your reality, just like magic! Also...MAGNET = IMAGE NET = IMAGE RETINA, where the word *retina* literally comes from the Latin, meaning *net*. Your MAGIC powers, which are magnetic, lie in your ability to imagine. [see the definition for PRINCE CHARMING]

MAGNET = IMAGINE IT. What you imagine, is what you attract into your life. You create your own reality through your imagination. Your imagination has a magnetic essence, which expresses through your pineal to project your reality upon the world. Your dreams come true through your imagination. Your fully functioning imagination is analogous to Prince Charming, your highest magnetic attraction. When you rise into the highest frequencies, you meet your dreams, and figuratively dance with Prince Charming. Open your imagination and dance with your dreams. [see the definition for IMAGINE IT and PRINCE CHARMING]

MASCULINE = MASK -*al* -*ine* = MUSCLE -*ine*. The masculine energy is the mask of illusion that establishes itself within your protein structure, and which strengthens with repetitive massaging. The masculine is simply the wounded feminine, as she dons her fiery reptilian mask, and flexes her angry muscles. [see the definition for MUSCLE]

MASTER MASON = 33rd DEGREE = 33rd DOWN STEP. Taking *33 steps down* the spinal-staircase, you arrive at the tip of the coccyx, where your first cell, your mother-of-all, your ovum, resides. At the end of your evolution, when you've reconnected to this point of power, the mother's loving heart, you will have MASTERED, or overcome, your petrified human body.

MATTER 1 = MATRE = MOTHER. Your body matter is designed from your very first cell, your ovum or egg...she is the *mother*-of-all-cells, both in size and duty. As such, she is the source of all creation. In her optimum state, she is the font of genius, the fountain of youth, the sparkling nest of infancy. When your matter is externally stimulated into action, she takes on her motherly role of defender and nurturer (immune system) as a means of protection. She protects her youth as a mother protects her infant. In this external world, your matter is always swaddling your infant design and keeping it safe behind blankets of emotion, until it can be freely expressed through your body, without threat of external reprisal. [see the definition for MOTHER]

MATTER 2 = M + ATTAR = E.M. ATTAR = **E**LECTRO **M**AGNETIC ATTAR = **E**LECTRO **M**AGNETIC SMELL. Attar is the perfume of flowers or essence, or in other words, smell (particularly rose essence). Your body matter is composed of cells containing electro-magnetic oceans of smell. [see the definition for OLFACTORY]

MEDITATION = IMMEDIATE *-ation* = the state of being present in the immediate moment. This is the gift of love that is always present, awaiting your choice. Being in the moment provides you with the opportunity to choose your thoughts, words and actions right now, instead of responding from pre-programmed triggers from the past. And when given this choice, who wouldn't choose love?

MEDIOCRITY. Moderation. The regulation of appropriate measures ensuring they satisfy *expected* parameters. Control. Government.

MEMBRANE = MEM + BRAIN = MA'AM BRAIN = FEMININE BRAIN. Cell membranes are also called plasma membranes. The word *plasma* means something that is molded or created, thus bestowing the title of *creative feminine brain* upon each and every cell membrane in your body. Your cell membranes are programmable, acting just like silicon chips. They hold information that can be altered to re-design your physical form. Their creative capacity is infinite.

MITOCHONDRIA 1 = MITO + CHONDRO = thread + cartilage. The mitochondria carry the threads of cartilage through the generations. Thread = yarn = story. Therefore, the mothers' stories are passed down through the generations within the skeleton. The fairytale of evolution is unfolding through the skeleton. Isn't it interesting that one of the definitions for skeleton is mummy?

MITOCHONDRIA 2 = MITO + CHONDRIA = thread, yarn or story + grain, little grain or grain of salt. The stories that are passed down via the mothers, need to be taken with a grain of salt. There's an element of untruth around the stories being passed down...they're patriarchal in nature, so need to be refined.

MITOCHONDRIA 3 = powerhouse, or generator of power = genius. Creative power is transported through evolution via your mitochondrial Mother Goose.

MOLECULE = L *moles* + *-cule* = barrier + diminutive = little barrier. Scientifically, a molecule is "the smallest part into which a substance can be divided without destroying its chemical character" (Oxford Dictionary of Etymology). Molecules, therefore, are little barriers of chemistry. Molecules are little barriers of smell/memory. Therefore, molecules are the barriers we're overcoming throughout evolution as we refine our characters; our behaviours, habits, customs; our memory; our smell.

MORTAL = MORTAR + *-al* = pertaining to MORTAR. Mortar is a cement mixture that's used in the building industry to bond bricks or stone, ensuring a rigid, inflexible structure. It consists of lime, sand and water, the exact components of the rigid, inflexible ossified skeleton (calcium, silica and water). The ossified skeleton, holding the shameful family secrets, is mortified!

MOTHER = MATRE = MATTER. The word *mother* not only means a nurturing female parent, but also a *source, origin* or *creator*. Your *mother*-nature, which is your matter, is the creator within the temple of your body. Your matter is the source from which you draw power, so listen to your body and tune in to what your matter is telling you by way of how you *feel*. [see the definition for MATTER]

MUSCLE = MASK *-al* = pertaining to your mask. Muscle is your fiery masculine; your wounded and angry feminine wearing her reptilian mask of defence. The mask of her persona expresses emotions through your muscles as they bully your body tissue into submission. [see the definition for MASCULINE]

MUTATION = the act or process of changing; and in the genetic sense, the process whereby heritable changes in DNA arise. (Online Dictionary of Etymology). Evolution is also the process of changing, including the progressive altering of DNA through the generations. Evolution is a mutation?

NERVE = N OEUVRE = NITROGENOUS OEUVRE = NITROGENOUS BODY OF WORK. This is the body you create by virtue of the messages you feed your brain via your nervous system, as dictated by the ancestral story carried within your nitrogenous DNA. You are the artist creator, and your physical body is the gallery within which your life's work is hung. A simple play on words provides a gentle twist in that the word *work* also means effort or toil, indicating that your nitrogenous body displays the wear and tear of your life's struggles through its aging tissue.

NOISE = NO EASE = DIS-EASE = DISEASE. Sound is simply a lower light frequency. It has a dualistic nature, for it requires one object hitting another in order to be heard. Sound, therefore, requires a barrier of some kind in order to exist; resistance is inherent. Sound creates dis-ease within your body tissue, contributing to its deterioration.

NORMAL. Conformity with rule or common standards. (Online Etymology) Being obedient to an *external* force. This requires following someone else's dictates and expectations. [see the definition for SOCIAL]

NURTURE = INERT *-ure*. When you soothe your fearful, nervous inner voices, you render them inert, or still. This is how the nervous system is being soothed and quieted throughout evolution. Every time you soothe a fear, you short circuit the automatic cycle that perpetuates your chemical reactivity (emotions). By doing so, you soothe your body tissue with the gentle higher frequencies that help to maintain its integrity; you employ the energies of creation rather than destruction. Nurturing your body tissue fosters youth within its structure and joy within its spirit.

OLFACTORY 1 = OLIVE FACTORY = OIL FACTORY. The olfactory system is the body's system of smell. It manufactures your essential oils; those oils that are essential to your body's functioning. As such, they contain your essence or smell, which is translated to your limbic system and stored in memory within the fabric of your being. Your olfactory system manufactures memories and stores it in your body matter in the form of emotions...creating your unique chemical signature. Hence, the chemical blueprint of your DNA is smell. [see the definition for MATTER, FUNCTION & CHROMOSOMES]

OLFACTORY 2 = ELF FACTORY = SANTA'S WORKSHOP, where the toys are made. A nosegay is an old-fashioned word for a toy, as well as a bouquet of flowers. So, the elf factory is where the floral scents (toys) are made, and SANTA'S WORKSHOP = SCENTER'S WORKSHOP = OIL FACTORY = OLFACTORY, which creates the body's scents/smells. With a further and unexpected connection to chromosomes, it becomes evident that the olfactory system manufactures smell from your DNA blueprint, which is imbued with your familial belief system. [see the definition for CHROMOSOME]

OPEN = UN-PEN. To release from an enclosure. Stepping out of the closet. To liberate or free from confinement, secrecy or imprisonment. Humanity being released from its evolutionary confinement within the cells of the body, having rehabilitated its lower behaviours. The opening of the third eye is the un-penning of the third ventricle from its defensive limbic borders.

PAIN = PAY N, with N being the Nitrogenous bases of your DNA. Pain, therefore, is the *cost* of following the old programmed family patterns (habits) that are inherent in your DNA. It's what ensues when your body bows down to your protein-packed, bossy muscles. Your feminine essence has given her power away to dis-ease. Your tissue is ill-at-ease because it's not operating according to the original design of your *mother-nature*. She's wounded, and she's crying out so you will know, and come to her rescue. Your wounded feminine has taken on a masculine edge in order to survive the family and society. [see the definition for MASCULINE & MUSCLE]

PASSION 1 literally means SUFFER = SUF + FER = CARRY UNDER. Your passion is the suffering of your true essence from its place of suppression, having deferred to the external world's expectations in the process of growing up. It's deeply wounded, feeling that it's been overlooked and undervalued. Its anger is the fire in your belly that drives you to express your truth. Your passion is your deepest desire; your fondest memory that wishes to be remembered and reinstated. Once your fondest memory or true essence is reinstated, the passion disappears.

PASSION 2 = PER ASH N = the ashes-to-ashes generational cycles are driven by the fiery passions of childhood suppression, your burning desires, that fester in your nitrogenous cells. Once all passion has disappeared, the evolutionary cycles will cease.

PERSONA = PER SONAR = PER SOUND = by way of the *lower* light frequencies we call sound. Your persona is the false mask you wear in body and character, which has been created by your inherited family belief patterns. You wear your persona in order to fit into society. However, the persona is evolving as it travels from generation to generation, as the demands of the external world soften. The mask of illusion is slowly dissolving to reveal the light of truth; your true creative essence that's waiting to express lovingly through your physical presence.

PHOENIX = A mythological bird, adorned with multi-coloured feathers. It's thought to symbolise rebirth, resurrection and renewal, due to its habit of rising renewed from the ashes of its predecessors after dying on the flames of a funeral fire. The multi-coloured nature of the phoenix hints at the rainbow of light within which humanity exists, otherwise known as the visible light spectrum. This also aligns with the rainbow of light that's associated with the energy chakras that sit along the spinal channel of each human being. The proverbial phoenix represents the cyclical nature of humanity throughout evolution, as each new generation rises from the ashes to ashes, and dust to dust of its ancestors. With each new cycle, DNA is revised, along with family beliefs and behaviours, thus providing an increasingly youthful evolutionary legacy. [see the definition for EMBRYO]

PINEAL: Bear with me, as this word is particularly tricky, but interesting. There are many hidden meanings tucked within the Bible parables, which actually relate to the evolution of the physical body, hence this first reference.

PINEAL 1 = PIONEER -*al*. Pertaining to a pioneer or forerunner, in this case, JOHN THE BAPTIST = JOHN/JEAN (French pronunciation)/GENE + BE APT -*ist* (to make apt is to adapt) = the GENE that ADAPTS = EPIGENETICS (the study of changes in organisms caused by modifying gene expression = refining or cleansing of belief and behaviour patterns = forgiveness = baptism within the cells = evolution). In the biblical parable, John the Baptist was the forerunner of Jesus, who practiced forgiveness and cleansing in preparation for the coming of the Light. The pineal reflects the epigenetic changes that result when we cleanse old belief patterns from our genetic make-up. As pioneer, the pineal leads the way into the unexplored territory of evolution, preparing us for the higher understandings that are to come. It is the portal into the field of light. [BAPTISE = cleanse, purify, christen or name = to make like Christ, who was *The Word made flesh = what you say, think and write... in other words what you believe... is what you become*]

PINEAL 2 = PER EYE N -*ial*. Pertaining to the nitrogenous information that's registered through the retina in the eye, and transported to the pineal via the hypothalamus. This nitrogenous information provides your dualistic human vision of lower light frequency. But, as explained in the above definition, it is constantly being evolved through epigenetic cleansing.

PINEAL 3 = PI N -*ial*. Pertaining to pi, the ratio of a circle's circumference to its diameter, provided by nitrogenous information (nitrogen cycle within the environment). The circle is the iris.

PINEAL 4 = PINE *-ial*. Pertaining to that for which one pines, yearns, desires or dreams. The pineal is the portal to the imagination, where our dreams come true, whatever they are...so be careful what you wish for. As you evolve your beliefs into a higher vision for yourself, you manifest a higher reality to match...your dreams come true. Your inner and outer realities unite. The *etymology* of the word PINE, being to *cause to starve*, *to languish* or *waste away*, aligns perfectly with the under-nourishing of the pineal gland through childhood, which leads to it being *consumed with grief or longing*. This is felt, from the child's perspective, as *punishment* or the *penalty* for growing up, which causes *pain*, *torment*, *affliction*, *suffering*, an *enduring penance*. (Online Etymology)

PINEAL 5 = PER EYE KNEEL. To kneel is to bend the knee or genuflect, especially in worship. The light that reaches the pineal is indicative of that which you worship; it's indicative of your beliefs.

PINEAL 6 = [see the definition for ATROPHY].

POLYPEPTIDES = POLY + PEP + TIDES = MANY + LIFE + TIMES. As proteins are habits, your protein polypeptides embody many life times of habit. Many life times of family tradition are held within your protein polypeptides. Your ancestors are flexing their muscles within the fibre of your being.

PRINCE CHARMING = PRIME MAGNETISM. Prince means prime, first or finest, so Prince Charming is your finest or highest magnetic essence that flows through your body to create your reality...your dreams come true. Prince Charming is your fully open imagination, from where you can dream any reality into existence. Prince Charming is analogous to the highest essence or finest flower...rose, which has the highest frequency. [see the definition for MAGNET]

PROTEIN 1 = PER ROUTINE or PER ROTE *-ine* = HABIT. Your proteins are simply patterns of belief that have been repetitively massaged; thoughts that you regularly revisit; frequent electrical firings between nerve cells; cells that have fired together and therefore wired together. Your patterns of belief, thoughts, electrical impulses, are responses to external stimulation. Your protein-filled body matter is a habit, and its instruction comes from the family traditions of your DNA; your rigid and traditional patriarchal programming. [see the definition for PROTEIN 2]

PROTEIN 2 = A protein is a polypeptide chain. POLYPEPTIDES = POLY + PEP + TIDES = MANY LIFE TIMES. Many life times of family tradition are held within your protein polypeptides. Your ancestors are flexing their muscles beyond the grave.

REALITY = REALTY. Creating your reality means manifesting that which exists in your life; that which is real; your personal realty.

REPTILIAN 1 = REPEAT *-ilian* (as in **repet**ition). The infant brain wearing its limbic mask of repetitive, instinctual behaviours represents the mythological fiery dragon protecting its treasure within the dark cave of your cells. All chemical reactivity is reptilian, including competitiveness, anger and defensiveness. Your reptilian is controlling, because it's scared. Ring any bells?

REPTILIAN 2 = ARCHIPALLIUM = PRIMITIVE CLOAK. Your original or primitive brain, the infant brain, is wearing its limbic cloak; the veil of time and limitation. PRIMITIVE = PRIME, FIRST, FINEST & PRINCE. So, the dreamy fairytale Prince Charming, who's always pressured by family obligation to marry for advantage, finds his physiological counterpart in the reptilian brain, which is your infant who's become cloaked by the duties of family responsibility. The magnetic infant qualities of love, happiness and youth epitomise your dreams and are, in fact, the dreamy prince charming that you're looking for, which is buried beneath the cloak of adulthood.

ROMANCE = AROMA -*ness*. Romance embodies chemical attraction. The body chemistry within your cells is your essence, your smell, your aroma. So, when you're romantically attracted to someone, your essence embraces their essence in an aromatic cellular hug. You really do unite on a cellular level. They become part of your story, as they nestle cosily into your limbic memory, and haunt you in your every waking moment. Breaking up is difficult, because their aroma lingers in your cells, making it hard to forget them. Romance is transported through evolution in the form of smell within your chromosomes. [see the definition for CHROMOSOME]

RUDOLPH = RUDE + OLF = RED + OLIVE/OIL (relating to the *olfactory/oil factory* system of smell). The shiny red nose of Rudolph represents the high-frequency smell of roses, or rose essence, that serves to guide the heart back to your loving *mother*-nature. Santa = centre or heart. Santa also = scenter, or that which provides scent. And an old-fashioned term for toys is nose-gays. So, the rosy essence of your *mother*-nature guides your heart as it delivers its loving bouquet into your body, lifting your spirits into joy. Your Christ light is born. [see the definition for OLFACTORY 1 & 2]

SARCOPHAGUS = FLESH + EATING = FLESH EATING. Limestone was reputed by the ancient Greeks to consume corpses and hence used for coffins. (Oxford Dictionary of Etymology) Therefore, a sarcophagus was considered to be a flesh-eating stone coffin. Limestone is primarily comprised of calcium carbonate, which is the active ingredient in lime, one of the three necessary components of your bony skeleton. As the calcium (lime) acts upon your silica framework, it turns your skeleton onto a flesh-eating sarcophagus...a bony skeleton. Interestingly, silica is used to make glass...Snow White's coffin!?

SKIN = ES/EX + KIN = OUT OF/FROM + FAMILY, TRIBE. Your skin is the most obvious expression of your DNA, or family patterns; your body tissue ('tis you) that clearly paints your life habits in unique tones and marks and textures and wrinkles. Your very essence is written upon the parchment of your skin, as dictated by the smelly concoctions in your little cellular oil pots.

SOCIAL = *f.* Latin sequor/sequi = TO FOLLOW. Autism, therefore, is characterised by difficulties in *following* the expectations of the mainstream world. Autistics lack the skills to *follow* the dictates of others. But autistics are forerunners in the evolutionary progression into higher intelligence, which places them ahead of mainstream understanding. Autistics are leaders...not followers.

SPINE = *f.* Latin for BACKBONE, THORN, PRICKLE. Figuratively, difficulties and perplexities.

TALKING CRICKET = GEMINI = twins = duality. Talking Cricket represents Pinocchio's conscience; the dualistic, finger-wagging, rule-keeping memories of the past that constantly vacillate between right and wrong. The instinctive ancestral voices of approval and disapproval that chirrup incessantly in your mind. Talking Cricket operates through duality, whereas the Turquoise Fairy operates through unity. Throughout the story, Pinocchio's critical inner voice is soothed into a gentle and nurturing inner guide.

TELEPATHY. From the Latin *tele* + *pathy* = feeling from afar. This happens most effectively on the higher frequencies. There's too much clatter bang on the lower, more dense frequencies of sound, which interrupts the communication light signals.

TERRA FIRMA = TERROR FIRMER = SCARED STIFF = PETRIFIED = MINERALISED = OSSIFIED. Terra firma is the solid ground or earth upon which we live. Your earthly existence is one of fear. This is reflected within the solid matter of your physical body, which takes its instruction from your nervous system (your system of fear). Your obedient cells transport these messages of fear to all your connective tissue, including your skeleton, creating mineralisation/ossification...and aging. Your body is petrified; scared stiff. This is indicative of the wicked stepmother (not your loving *mother-nature*) losing her youth and beauty.

THIRD EYE = A mystical invisible eye, otherwise known as the mind's eye or inner eye, which engages a heightened perception supposedly beyond that of ordinary sight. The brain's third ventricle is an inner cavity that is continuous with the spinal channel, and which is located directly between the brows of the forehead, in perfect alignment with the mystical third eye location. [see the definition for THIRD VENTRICLE]

THIRD VENTRICLE: The third ventricle is a cavity within the mammalian brain, which is filled with fluid in its lower vibratory, human form. Being continuous with the spinal channel, it connects with the flow of information that courses through the neuro-endocrine system of the human body. However, in its high vibratory, original design, the cavity of the third ventricle is filled with light, which flows through the open spinal channel (hyaline canal) from the higher power source that's situated at the base of the spine...the original cell or ovum/egg. [see the definition for WINDOW and THIRD EYE]

TIME = TIE M = TIE E.M. = TIE ELECTRO MAGNETISM = E_M

TIME: TIE M = TIE E.M. = TIE ELECTRO MAGNETISM. The Limbic System links the Neuro-Endocrine systems within the brain. The neuro, or nervous system, is the body's electrical system **(E)**, and the endocrine, or chemical system, is the body's magnetic system **(M)**. Therefore, in linking the Neuro-Endocrine systems, the limbic system *ties* the body's *electro-magnetic* systems together, and in so doing it records TIME within the human body. It records both the mammalian progression through evolution, and aging within the individual. The electro-magnetic oceans within the cells are your unique emotional signature [see the definition for EMOTION]. Accordingly, TIME is measured by EMOTION. The limbic ties are the ties that bind you to the old belief and behaviour patterns that are stored in memory. The family ties that ensure your compliance with the external world and all its expectations.

TISSUE = 'TIS YOU. *'Tis you* who fills your body tissues with your unique energy and essence. It's *your* story that designs, creates and evolves every tissue in your body; every fibre of your being. 'Tis you that radiates from *your* body, creating *your* world. 'Tis you who determines your body's form. 'Tis you who determines the functioning of your body. 'Tis you!

TRADITION = a doublet of TREASON, meaning to BETRAY. The word tradition literally means to *hand over*, and refers particularly to the handing down of practices or beliefs based on Mosaic law. In reality, Mosaic law is analogous to the family code of conduct that is *set in stone* within the double helix of DNA, and handed down through the ancestral line. To betray is to violate by unfaithfulness. The infant is unable to remain faithful to their true design of youth and infinite possibilities when the limiting set of family beliefs is thrust upon them. Their innocence is violated...they are betrayed...by tradition.

TRUST = TRU *-ist* = TRUEST. The words trust and truth are cross-referenced within dictionaries, as they're related through the same Latin derivative. Truth denotes integrity, which is the truest representation. When a fabric has integrity, it holds true to its original design. It remains faithful to its original condition. Faith is trust. When you're true to yourself, you are faithful to your beliefs; you maintain your integrity; you have built-in trust.

TRUTH = TRU -eth = TRUE WORTH. Your true worth lay deep inside you, sparkling and crystal clear, wishing to be expressed. It's your pure light of truth that's radiant with love. When you love yourself, your true worth appears.

TURQUOISE FAIRY = INTUITION = INNER TUITION from your INNER TUTOR or guide. Your intuition is your loving inner voice, which soothes and nurtures you with kindly thoughts and words. [see the definition for INTUITION]

UNDERSTANDING = That which STANDS UNDER you, as the foundation from which you express; your bank of beliefs; the stories you tell yourself; your well of creative offerings; your imagination.

WATER: The obvious definition refers to the transparent liquid that is so crucial to our survival, and which fills our seas, lakes, rivers and oceans etc. The second definition is a lovely surprise, for it tells us that water is also the transparency and lustre of a gem...in other words its clarity and light or brilliance. Whereas the *fire* of a gem is evident in the flashes of twinkling rainbow lights on its surface, the *water* sits right at the heart of the gem indicating its truth and purity. (Concise Oxford Dictionary of Etymology)

WINDOW = WIND EYE = a VENT through which wind (and light) passes. (See the definition for THIRD EYE & THIRD VENTRICLE)

WOODS: Snow White is sent deep into the dark woods, which is analogous to your pure essence being sent deep into your dark cells to create your body matter. **Matter** not only derives from the French meaning *substance*, but also from Latin meaning the *hard inner wood of a tree*, and from Greek meaning *wood*. Matter also refers to the source or origin, and to mother. So, your *mother*-nature (your body matter) is protecting you from the external onslaughts by swaddling you in a thick blanket of emotion deep within your body matter. Her intention is to keep you safe, until you find your own sweet, loving voice, and no longer need to hide. (see the definition for MATTER]

YOUTH = YOU -*th*. = YOU, in your true state, embody YOUTH. Youth is simply your body tissue expressing integrity. You nurture that integrity by maintaining a constant flow of creative energy in every thought, word and action. For creative energy creates, whereas destructive energy destroys the fabric of your being, making it age. In infancy, your first and fondest memory was of your body tissue flowing with creative energy and love. Delta brainwaves were your natural pattern, as you were still connected to your fully open imagination and intuition, filling your tissues with the creative energies that constantly nurtured your body, regenerating its youth. [see the definition for DELTA, INFANT and BEAUTY]

Bibliography

Singh, V. (2012) *Consciousness and The Third Eye*. Speaking Tree Blog.

https://bit.ly/4cTPOCX

Cummings, E.E. American poet.

Einstein, A. *Quote*.

Twain, M. *Quote*.

Einstein, A. *Quote*.

The Imitation Game. (2014) *Movie*. The Weinstein Company.

Evolution. Online Dictionary of Etymology.

Swift, J. (1706) *'Thoughts on Various Subjects, Moral and Diverting'*

Sheen, Archbishop F.J. *Quote*.

Barnett K. (2014) *The Spark: A Mother's Story of Nurturing Genius and Autism*, (page 250).

Picasso P. *Quote*.

Life Coach Code. (2017) *New Research Shocks Scientists: Human Emotion Physically Shapes Reality*.

Suskind R. (2016) *Life Animated, movie. (2014) Life, Animated; A Story of Sidekicks, Heroes and Autism* (book).

Higashida N. (2014) *The Reason I Jump*

Barnett K. (2014) *The Spark: A Mother's Story of Nurturing Genius and Autism* (page 249)

Bukowsky, Charles. American poet and novelist (1920 – 1994) *Quote.*

Magnetic Sixth Sense. Maverick Scientist thinks he has discovered Magnetic Sixth Sense. *Science.org.* (2016) https://bit.ly/3UgKwdl

Baby's Brain Begins Now: Conception to Age 3. *Urban Child Institute.* https://bit.ly/3xFqMra

Interview with Dr. Temple Grandin. for Synapse-reconnecting lives. https://www.autism-help.org/story-temple-grandin-autism.htm

Interaction Design Foundation. *Our Three Brains - The Emotional Brain.* https://bit.ly/3UhI8mJ

Persona. Oxford Dictionary of Etymology.

Lancer, Darlene, JD, LMFT. *The Co-Dependent False Self*

Wikipedia, Limbic System. https://en.wikipedia.org/wiki/limbic_system

Emerson, Ralph Waldo. *Quote.*

Einstein, Albert. *Quote.*

Heritability of Autism. https://en.wikipedia.org/wiki/Heritability_of_autism

Dawkins, Richard (1976). *The Selfish Gene*

Wikipedia, Meme. https://en.wikipedia.org/wiki/Meme

Kong, Henry (2006) *More Self Than Self: At Autism's Edge* (page 203)

De Braganza, Caroline (2019) Revealing the Truth About Our Memory Banks.

Collu, R. et al. *The Pineal Gland - A Neuroendocrine Transducer.* https://bit.ly/3JjNh7t

Pineal. www.sciencedirect.com/topics/neuroscience/pineal-recess.

Grandin, Temple (1995) *Thinking In Pictures, (Chapter One)*

Murphy, Andye (2020) *The Pineal Gland and the Third Eye Chakra, Gaia* https://www.gaia.com/article/pineal-third-eye-chakra

Magnetic Sixth Sense. Maverick Scientist thinks he has discovered Magnetic Sixth Sense. *Science.org.* (2016) https://bit.ly/3UgKwdl

Shomrat, T. Nesher, N. (2019) *Updated View on the Relation of the Pineal Gland to Autism Spectrum Disorders.* https://www.ncbi.nlm.nih.gov/pmc/articles/PMC6370651

Homeostasis. https://en.wikipedia/wiki/Homeostasis.

Booth, Frances. M. FRACS, DO (Lond) (1987) *The Human Pineal Gland: A Review of the Third Eye and the Effect of Light.*

Picasso, Pablo. Quote.

Walsh, Colleen (2020) *What The Nose Knows.* https://bit.ly/3W5FRMu

Whitehouse, Professor Andrew (2018) *National Guidline for Assessment & Diagnosis of ASD in Australia*

Einstein, Albert. Quote

Mazzeo, Tilar J (2010) *The Secret of Chanel No. 5*

Zaraska, Marta (2017) *The Smell of Scent in Humans is More Powerful than we Think.* Discovery Magazine. https://bit.ly/3VWbHeI

Autism Spectrum. https://en.wikipedia.org/wiki/Autism_spectrum.

Einstein, Albert. *Quote.*

Shakespeare, William (1603) *Othello.*

Shah, Perita (2020). *A Primer of the Chakra System.* Chopra.
https://chopra.com/blogs/meditation/a-primer-of-the-chakra-system

Oleson, Jacob (2013) *Colour Meanings: The 7 Chakras and their Meanings.*

https://bit.ly/3VZsQnR

Woolf, Virginia (1925) *Mrs Dalloway.*

Silica: A Little Known Element Comes of Age. Eidon Ionic Minerals.
https://www.eidon.com/silica_article.html#:~:text=Bone

Collagen. https://en.wikipedia.org/wiki/Collagen.

Rose Oil. www.aromaticamedica.tripod.com/id19html

Black, Sue (2020) *Written in Bone: hidden stories in what we leave behind* (page 232)

Einstein, Albert. Quotes.

Black, Sue (2020) *Written In Bone: hidden stories in what we leave behind*

Kreiger, Nancy (American epidemiologist) from Black, Sue (2020) *Written In Bone: Hidden Stories in what we Leave Behind* (page 327)

Harari, Yuval Noah (2011) *Sapiens: A Brief History of Humankind.* (pages 30-31, 40, 43)

Mitochondria: *National Human Genome Research Institute.*
https://www.genome.gov/genetics-glossary/Mitochondria (updated 2024)

Zuidena, Dalen et al (2023) *Identification of candidate mitochondrial inheritance determinants using the mammalian cell-free system.*
https://pubmed.ncbi.nlm.nih.gov/37470242/

Sykes, Bryan (2003) *Adam's Curse: A Future Without Men*

Chatterjee, Das et al. (2022) *Mitochondrial Epigenetics Regulating Inflammation in Cancer and Aging.* National Institute of Health.
https://pubmed.ncbi.nlm.nih.gov/35903542/

What Is Epigenetics? (2022) *Centres for Disease Control and Prevention: Genomics and Precision Health.*
https://www.cdc.gov/genomics/disease/epigenetics.htm

Mandl, Mike. 2020. *Meridians, Maps of the Soul.*

Lewis, C.S. Quote from Callahan, Patti (2008) *Becoming Mrs. Lewis*

Fairchild, Alana (2017) *Love Your Inner Goddess*

Proteins: building blocks of the body. *Otsuka Pharmaceutical Co, Ltd*
https://bit.ly/4aRmISO

Rumi, Jalaluddin. Quote.

Michigami, Toshimi & Ozono, Keiichi. (2019) *Roles of Phosphate in Skeleton.* NIH. PubMed.

Roth, Gabrielle. Quote.

Gray, Henry *Gray's Anatomy"* Hyaline Canal (page 191)

Kundalini; the serpent power The Satva Collection. (2023)
https://www.thesattvacollection.com/blogs/news/kundalini-the-serpent-power

Gupta, N et al. (2017) *Disorders of neural tube dev't* Filum Terminale - an overview. https://bit.ly/43Z96CM

Lipton, Bruce (2005) *The Biology of Belief: unleashing the power of consciousness, matter and miracles.* (page 87)

Lance, Darlene JD, LMFT *The Co-Dependent False Self*

Einstein, Albert. *Quote.*

Wilson, Ruth (2022) The Jane Austen Remedy (page 296)

Craft, Samantha, aka Ciampi, Marcelle MEd. *Quote.*

Spalding, Baird T. *Life and Teaching of the Masters of the Far East.* (page 51)

© Marg Kinneen 2018 - 202

About the Author

Marg Kinneen was inspired to offer an alternative perspective about autism after her Aspergers diagnosis in 2016 and the experiences that followed.

She has a background in teaching and stage performance, and therefore has always been a presenter of information.

Coupled with forty years of investigation into spirituality, energy medicine and new thought, she now straddles the arenas of spirit and science, as she re-imagines human spiritual and physical evolution from a cellular perspective.

Conscious evolution has always been her platform, and fairy tales have always been her fascination.

In *Autism & Evolution: Spirit, Science & Fairy Tales*, Marg presents a unique and profound idea, along with a heartfelt wish to awaken this sleepy world to the truth about autism.

www.ingramcontent.com/pod-product-compliance
Lightning Source LLC
Chambersburg PA
CBHW030312080526
44584CB00012B/532